SATELLITE TELEVISION

Satellite Television: Techniques of Analogue and Digital Television

H. Benoit

A member of the Hodder Headline Group
LONDON • SYDNEY • AUCKLAND
Copublished in North, Central and South America by
John Wiley & Sons Inc., New York • Toronto

First published in Great Britain in 1999 by Arnold,
a member of the Hodder Headline Group
338 Euston Road, London NW1 3BH

http://www.arnoldpublishers.com

This book is adapted and translated by the author from Benoit, H., 1998:
Télévision par satellite: Technique de la réception analogique et numérique.
Paris, Dunod.

Copublished in North, Central and South America by
John Wiley & Sons Inc., 605 Third Avenue,
New York, NY 10158–0012

British Library Cataloguing in Publication Data
A catalogue record for this book is available from the British Library

Library of Congress Cataloging-in-Publication Data
A catalog record for this book is available from the Library of Congress

ISBN 0 340 74108 2
ISBN 0 471 35824 X (Wiley)

1 2 3 4 5 6 7 8 9 10

Commissioning Editor: Sian Jones
Production Editor: James Rabson
Production Controller: Sarah Kett
Cover Design: Stefan Brazzo

Typeset in 11/12pt Times by J&L Composition Ltd, Filey, North Yorkshire
Printed an bound in Great Britain by J.W. Arrowsmith, Bristol

Contents

Preface

Satellite television started in Europe with some knowledgeable and relatively wealthy amateurs in the first half of the 1980s, but its democratization really started in 1989 after the launch of the first ASTRA satellite (ASTRA 1A) at the orbital position of 19.2°W.

It is now part of the daily life of millions of European television viewers, and it is now possible to find in specialized shops or department stores a complete analogue system (fixed antenna plus receiver and cables) for less than UK£100 (approximately 150 euros).

The analogue form of satellite TV did not, however, follow the same development scheme in all European countries, it depended on some political–technical choices made in the 1980s (high power satellites like TDF 1/2 and TV-SAT, D2MAC etc.) and the relative programmes available on terrestrial and satellite television.

In the UK and Germany, analogue satellite TV developed very quickly from 1990 onwards, mainly from the 19.2°E orbital position, but with different approaches: for the UK, programmes offered were mostly pay-TV (mainly from BSkyB) whereas for Germany the programmes were practically 100% 'free-to-air' (of the order of 20 programmes). Both approaches resulted in many millions of installed receivers in both countries in the mid-1990s. This was not the case in France, which strongly pushed D2MAC and high power TDF satellites and, later, Telecom 2A/2B, with a very limited programme, and an even more limited success.

It is probably for this reason that France was the first European country to start digital satellite television services in 1996 (Canal+/Canal Satellite Numérique on ASTRA 19.2°E, AB-SAT and TPS on Eutelsat 13°E). The success of these digital 'bouquets' quickly exceeded the most optimistic forecasts.

However, the launch of digital satellite TV in Germany (DF1 in 1997) was very disappointing, and the UK digital satellite television service (BSkyB) had just been launched at the time of writing.

The objective of this book is to present, in a simple and concise manner, all the aspects of satellite TV reception that are necessary to understand its operation and make the best use of a receiving installation, be it analogue or digital.

This book's target readership is the professional or the student desiring to acquire or update knowledge in this field, as well the amateur who wants to set up his or her own installation, improve it or simply understand 'how it works'.

Prerequisite knowledge of the subject is not important: although undergraduate level mathematics would be useful for some (non-essential) calculations, and knowledge of basic analogue and digital electronics would be helpful for the receiver explanations.

1 Principles of satellite television

1.1 Short historical reminder

Radio waves of very high frequency (above 50 MHz), which are necessary to carry the high bandwidth of TV signals, have the property that they do not propagate far beyond the horizon, and this becomes worse when their frequency increases (and thus their wavelength decreases), since their propagation becomes very similar to the propagation of light.

This is not the case for radio waves of greater wavelength, as used for AM radio (especially for medium and short wave frequency bands, which vary between 500 kHz and 30 MHz), which in some conditions are reflected by the ionized upper layers of the atmosphere, thus allowing very long distance communications. It has to be admitted, however, that such communications are not always very reliable.

In order to increase the range of ultra-short-wave communications, various active or passive reflectors placed at high altitude (reflecting balloons in the high atmosphere or a relay placed on an aeroplane that is turning in a restricted area for retransmission of a special event) have been used. Even the Moon has been used as a reflector for experimental purposes!

The first transatlantic 'Worldvision' TV transmissions (USA/UK and USA/France) were made possible thanks to the Telstar 1 satellite in July 1962.

This satellite and its immediate successors had, however, the considerable drawback of being placed in a low orbit (some hundreds of kilometres). Their rotation, therefore, was much quicker than the Earth's rotation.

This allowed only short transmission 'windows' of some 20 or 30 minutes during which the transmitting and receiving stations had to track and follow the satellite using their enormous horn or parabolic antennas. These antennas often had to be placed under a 'radome' to protect them against the atmospheric conditions. In Europe, it was the stations of Pleumeur-Bodou in France (located in Britanny, near Lannion) and of Goonhilly Down in England that had the privilege of making possible these first transatlantic links.

The drawbacks of these first non-permanent links were finally overcome in August 1964 thanks to the launch (after two failures) of Syncom 3, the first 'fixed' telecommunications satellite. Syncom 3 was the first satellite to be successfully placed in a *geostationary* orbit, allowing it to be seen at a fixed position from the earth.

The era of global telecommunications had really started.

1.2 The geostationary orbit

The idea to place satellites in a geostationary orbit is not new: it was the English writer and engineer Arthur C. Clarke who first proposed, in an article in *Wireless World* (published in 1945), a global radiocommunication coverage by means of three 'fixed' satellites situated 120° apart on a circular orbit at approximately 36 000 km above the equator (Fig 1.1).

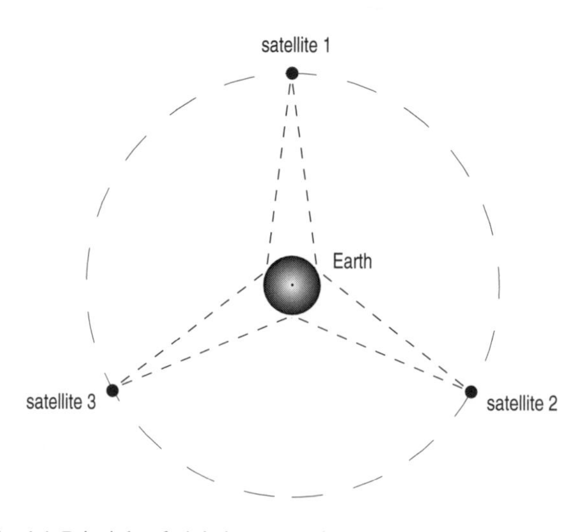

Fig. 1.1 Principle of global coverage by three geostationary satellites

This project was unrealizable and could even have seemed somewhat crazy at the time; however, it was less than 20 years before such a realization began (with Syncom 3), and its concept was very near that which was put in place 40 years later by the international organization INMARSAT for long distance communications with ships.

The geostationary orbit is unique for the following reasons.

- In order that the satellite appears at a fixed position relative to the Earth, it must turn on a fixed orbit around the Earth, and at the same angular speed. Since the Earth revolves around its polar axis, this orbit must be circular and centred on the polar axis.
- In order for this orbit to be fixed, the centrifugal force due to the satellite rotation must be exactly compensated by the terrestrial attraction, so these two forces have to be collinear and opposed. This can only happen in the equatorial plane, where the orbit must then be situated.
- In order for the angular rotation speed of the satellite to be exactly the same as the rotational speed of the Earth (one revolution every 23 hours 56 minutes, which is the duration of the sidereal day), calculation shows that the radius of the circular orbit must be 42 200 km. Taking into account the radius of the Earth (approximately 6400 km), this means that the geostationary orbit is located at an average altitude of 35 800 km above the equator.

The placement of a satellite on a geostationary orbit requires two main steps:

- *launching*, by means of a rocket (Ariane, Proton, Long March, etc.) or by the Space Shuttle, which places the satellite in an elliptical equatorial orbit with a *perigee* of some hundreds of kilometres and an *apogee* of 35 800 km;
- *placement in position*, which is performed, starting from the apogee, by correcting the satellite's orbit by means of its on-board motors, in order to make the orbit circular, and positioning the satellite in the particular place assigned for its mission. The amount of propellant used for this phase, which depends on the launching accuracy, is critical since it influences the operational lifetime of the satellite. This is due to the fact that, during the operational life of the satellite, in order to maintain its position within an apparent angle of less than 0.2° as seen from the Earth (corresponding to a volume of approximately 140 km diagonal) it will be necessary continuously to make

minor corrections by means of the on-board motors. Therefore, except in case of functional faults, it is mainly the remaining quantity of propellant after initial positioning that determines the useful life of the satellite.

The main on-board energy source is for the electronic equipment of the satellite (transponders) and is produced by large solar panels. In the most recent satellites, these can continuously produce more than 5 kWh of electrical power. They are also buffered by batteries that can ensure partial or total operational autonomy for the duration of lighting eclipses which occur in the equinox periods (see section 1.2.2).

1.2.1 Azimuth and elevation of geostationary satellites

Let us first remind ourselves of the concepts of *latitude* and *longitude*. In order to locate a point on the Earth, the globe must be graduated with imaginary circles (see Fig. 1.2).

- *Parallels* are circles that are parallel to the equator; their position, or latitude, is defined by the angle between the equatorial plane and the line joining the particular parallel to the centre of the Earth; it has a value between 0 and 90°.

  The latitude of a place is expressed in degrees of latitude North or South depending upon which hemisphere is being considered.

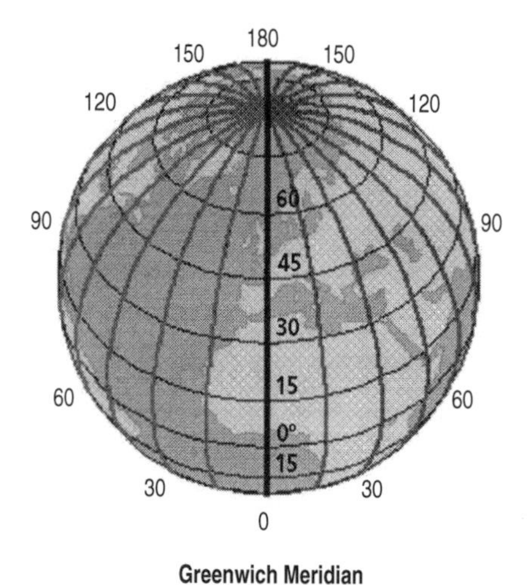

Fig. 1.2 Latitude and longitude, meridians and parallels

One degree of latitude corresponds to approximately 111 km of North–South displacement. For Europe (including Scandinavia), latitude varies between 30° and 60° of latitude North.

- *Meridians* are imaginary circles passing through the poles; their position, or longitude, is defined by the angle between their plane and the plane of a reference meridian (named the *Greenwich Meridian*) which passes through a former observatory located in a suburb of London.

  The longitude of a place is expressed in degrees and varies between 0 and 180° East or West of the Greenwich Meridian, depending on its relative position. For Europe, longitude varies between approximately 10° West (Portugal) and 30° East (Greece).

Seen from a point on the Earth, the position of a geostationary satellite can be defined by two angles.

- *Elevation*: the angle between the imaginary line joining the place of reception to the satellite and the ground (the horizontal plane at the place of reception).
- *Azimuth*: the angle between the horizontal projection of the station–satellite imaginary line and the plane of the meridian (North–South) passing through the station. This angle is measured from the North (so the azimuth of full South of the station is 180°)—see Figs 1.3 and 1.4.

The position of geostationary satellites is defined by their longitude (relative to the Greenwich Meridian, so relative to this meridian the azimuth of the satellite is the same as its longitude); since such satellites are all located in the equatorial plane, their latitude is of course 0°, and therefore does not need to be mentioned. For an observer situated in the Northern hemisphere, all geostationary satellites are 'seen' on an arc centred on the direction of the South Pole, of which the 'height' varies depending on the latitude of the location (see Fig. 1.5).

- The elevation of a satellite situated full South (azimuth 180°) depends directly on the latitude: it is approximately 48° in the South of Greece (lat. 35°N) and 11° in the North of Norway (lat. 70°N).
- The azimuthal range of satellites that are theoretically 'visible' (i.e. situated above the horizon) depends on the latitude; at a latitude of 45° it ranges approximately from 70° East to 70° West.

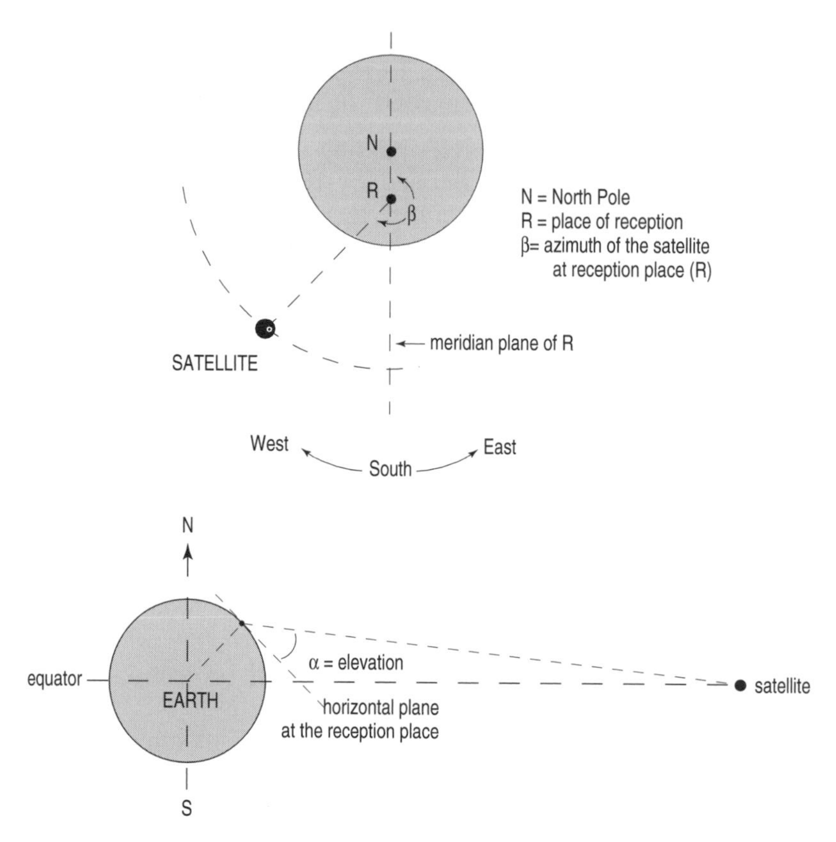

Fig. 1.3 Azimuth and elevation of the satellite at the place of reception (seen from space)

1.2.2 Equinox eclipses

Due to the geostationary orbit being in the equatorial plane, a part of the orbit is in the shadow of the Earth (see Fig. 1.6) for two periods of approximately 45 days, centred on the Spring and Autumn equinoxes (21 March and 21 September); effectively, at the equinox, the Sun's orbit is exactly in the equatorial plane, so there is—for any geostationary satellite—a moment in the day when it is in the shadow of the Earth.

For a satellite located at 'full South' of the place of reception (azimuth 0°), this lighting *eclipse* is centred around midnight (local solar time) and has a maximum duration of 70 minutes on the day of the equinox.

When the satellite is in the Earth's shadow, the supply of electrical energy normally delivered by the solar panels is completely

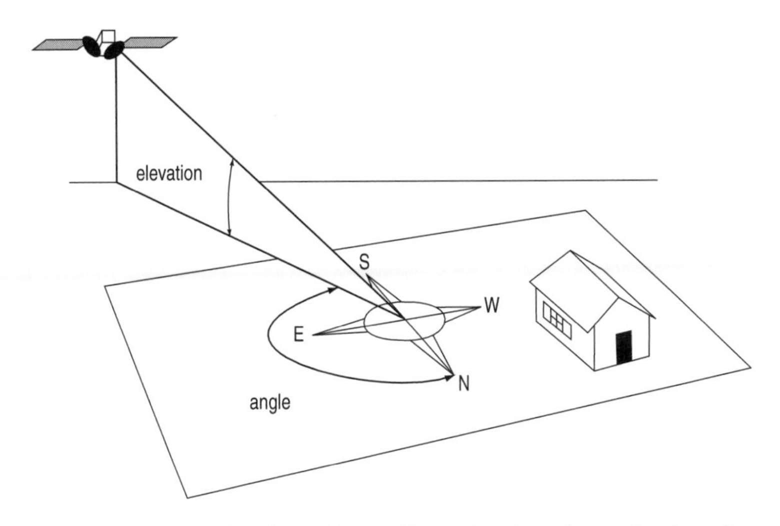

Fig. 1.4 Azimuth and elevation of the satellite at the place of reception (seen from the Earth)

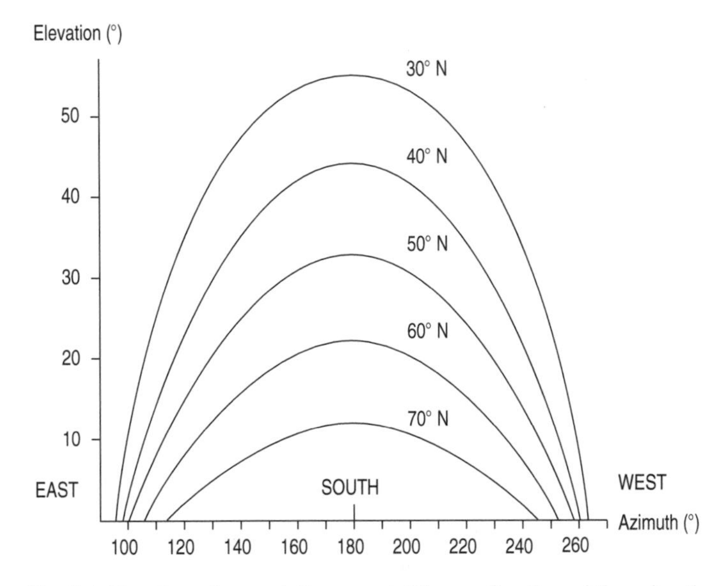

Fig. 1.5 Elevation of a geostationary satellite as a function of the azimuth

interrupted, and, during the time of the eclipse, the satellite has to rely on its batteries for operation. This is why the position allocated to the direct broadcast satellites (WARC 77 plane) was offset to the West compared with the service area, so that the eclipse happened very late (between 2 and 4 o'clock in the

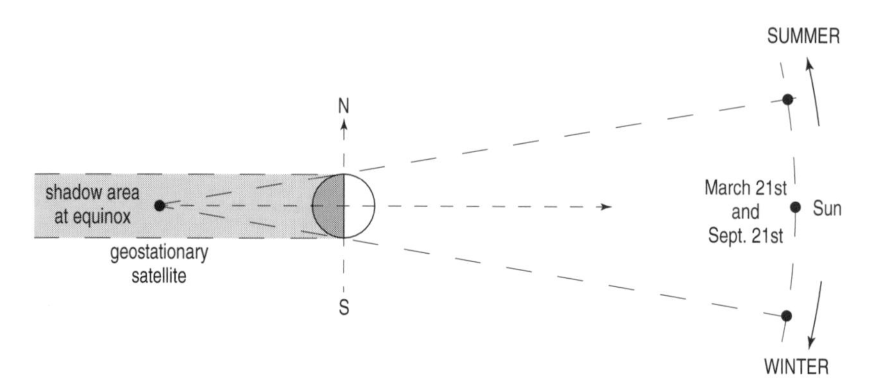

Fig. 1.6 Sunlight eclipse around the equinox periods

morning); this would allow only a partial interruption of the service with minimal disturbance for the users.

Table 1.1 gives the duration of the solar eclipse as a function of the distance to the equinox (in days before or after the equinox).

Table 1.1 Eclipse duration as a function of the distance to equinox (before and after)

Distance (days)	23	22	20	16	12	10	0
Duration (min)	0	10	35	50	60	66	73

Outside these periods, the solar orbit makes a sufficient angle with the equatorial plane to ensure that the satellite is never in the Earth's shadow.

There are also eclipses due to the passage of the Moon between the Sun and the satellite; they have the same effect, but their occurrence is much more infrequent.

1.2.3 Outage of the reception antenna

There are also two periods of a few days every year (centred two or three weeks before the Spring equinox and after the Autumn equinox in Europe), during which the Sun is in the alignment of the satellite for a short time (see Fig. 1.7).

Since the Sun is a very powerful wide-band electromagnetic noise generator, it can degrade the signal-to-noise ratio to the point where reception can become very disturbed or even impossible. This 'outage' can be considered as a kind of 'dazzling' of the low noise block (**LNB**) converter by a very strong signal. This is one of the reasons why telecommunication stations generally use

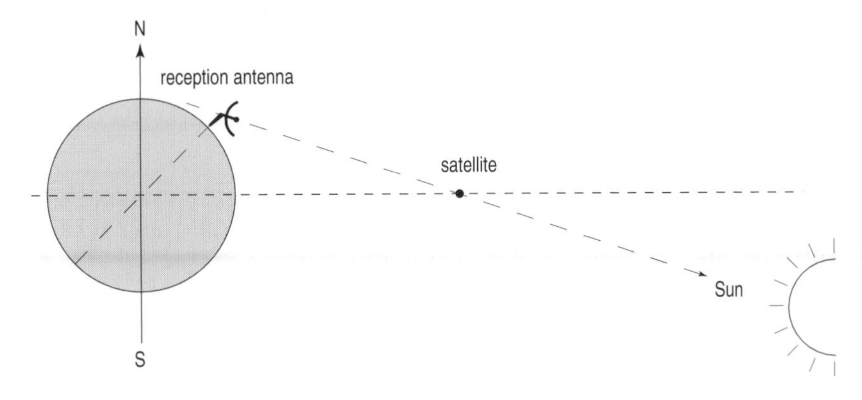

Fig. 1.7 Antenna 'dazzling' by solar radiation

two sites, sufficiently separated such that this phenomenon occurs at different times for each station (this is also useful in the case of very strong atmospheric disturbances, which are not very likely to occur simultaneously at both places). The duration and intensity of this dazzling, which takes place around noon (local solar time) for a satellite located at 'full South', depends on the directionality, and thus on the size of the antenna. The duration of this dazzling (number of minutes per day and number of days affected) is inversely proportional to the size of the antenna due to the increase in directionality with size. However, the intensity of disturbances increases with size due to the increase in gain with size (Table 1.2). For instance, a 50 cm antenna can be affected for up to 15 minutes per day during approximately 10 days, whereas a 1 m antenna will be affected for only 8 minutes during 6 days, but more strongly.

Table 1.2 Duration of dazzling outage as a function of antenna diameter

Φ (cm)	40	50	60	70	80	90	100
Dir. −3 dB (°)	4.5	3.6	3.0	2.5	2.2	2.0	1.8
Days affected	18	15	13	11	10	9	8
Duration (min)	14	11	9	8	7	6	5

1.3 Frequency bands and polarization

Apart from their use for military transmissions, the first geostationary satellites were used mainly for telecommunication purposes such as intercontinental telephony and occasional or permanent links between television and radio broadcasting organizations.

1.3.1 Frequency bands

The choice of frequency bands used for satellite communications has been a compromise between the advantages and drawbacks of different usable bands as well as the pre-existing frequency allocations and the technological possibilities of the moment.

Mainly due to the technological limitations and cost considerations for the receivers, the frequencies used initially had to be relatively low (below 4 GHz). However, important studies have been carried out by international telecommunications organizations to determine which frequency bands were best suited for geostationary satellite services.

Two main parameters, often contradictory, have to be taken into account.

1. *Noise level (natural or man made)*
   Except for the background noise due to the 'Big Bang', which is constant up to more than 500 GHz, all other sources of noise (atmospheric, galactic or artificial) decrease almost linearly with frequency, and become practically negligible above 5 GHz.

2. *Attenuation due to the atmosphere and atmospheric disturbances (mainly rain, snow and fog)*
   Atmospheric attenuation by clear weather is low below 15 GHz and then increases with frequency, with some remarkable peaks, notably the one due to absorption by water (around 22 GHz) and the other much more important one, that due to absorption by oxygen (around 60 GHz), which makes the atmosphere practically opaque for this frequency range (attenuation of around 10 dB/km). Concerning the attenuation due to rain, it is very limited below 3 GHz and increases progressively up to 80 GHz, where it reaches an almost stable maximum. Thus, the most appropriate frequencies seem to be between approximately 3 and 5 GHz (Fig. 1.8). This is the main reason why the first band used is the so-called 'C band' (from 3.7 to 4.2 GHz), another reason being the increasing technological difficulties with higher frequencies.

However, another important element to be taken into account (especially for 'direct-to-home' services) is the size of the antenna, because performances (in term of gain and directivity) increase with frequency for a given dimension (see Chapter 3). This is why

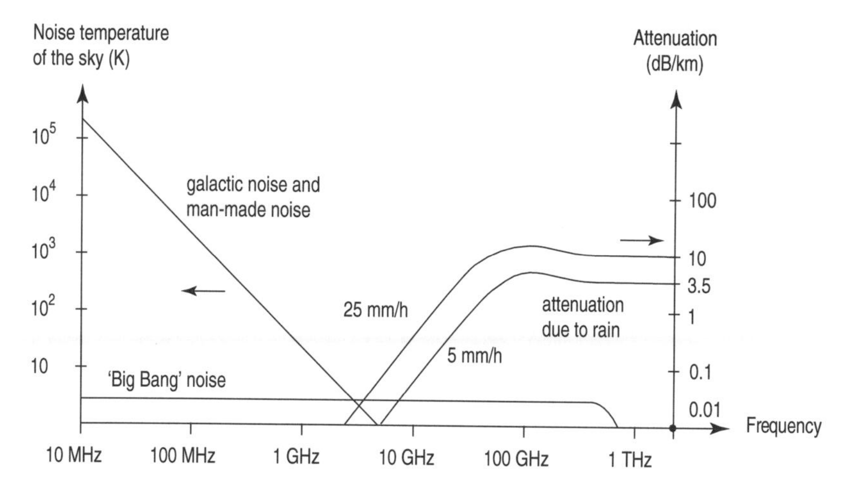

Fig. 1.8 Noise temperature and main attenuation as a function of frequency

the Ku band (10 to 15 GHz approximately) has been preferred for telecommunications in 'temperate' countries such as Europe (with few very heavy rainfalls) (**FSS**—Fixed Satellite Service—uses bands from 10.950 to 11.700 and 12.500 to 12.750 GHz for what is called the *downlink* path).

Meteorological studies have allowed us to model the statistical distribution of rain intensity (measured in mm/hour) in different parts of the world, divided into five regions. These studies have shown that, for Europe (region 3), a rain intensity of 5 mm/hour is only exceeded for 0.25% of the time in one year, which corresponds to 1% of the time of the worst month. So, if one wants to guarantee reception for 99% of the worst month (which is the figure generally used to calculate the installation), it is implicit that this rain intensity of 5 mm/hour be taken into account.

At 12 GHz, the attenuation corresponding to this rain intensity is approximately 0.2 dB/km. The effective resulting attenuation depends on the thickness of the rain layer (i.e. the layer in which it is raining), through which the signal has to be transmitted, and then on the elevation of the antenna: at a given rain intensity, attenuation will become increasingly important as the elevation decreases (see Fig. 1.9).

From Fig. 1.9, one can see that for an elevation of 30° (average value in mid-Europe) this attenuation is of the order of 1.5 dB. This value can be considered as a maximum value (not exceeded for 99% of the worst month, which corresponds to 99.75% of the whole year).

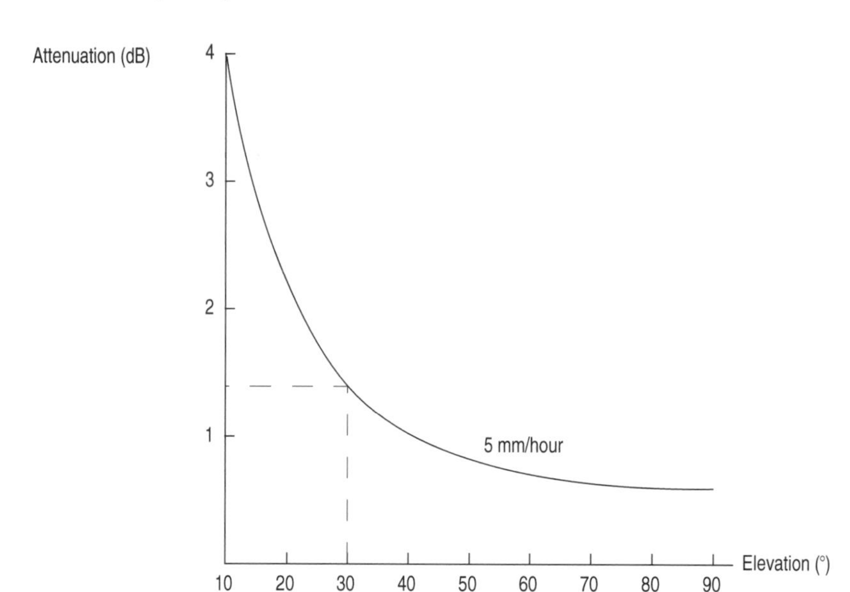

Fig. 1.9 Rain attenuation (5 mm/h) as a function of antenna elevation at 12 GHz

It can happen, however, during short periods of exceptionally strong rains or snowfalls, that the attenuation will be much higher, up to the point where reception is very degraded or even impossible. If one wants to reduce these periods to the very minimum, then one should, for instance, ensure reception during 99.9% of the worst month—which corresponds to a rain intensity of 25 mm/hour, and which results in a much stronger attenuation and therefore a much bigger antenna size. However, this is normally not required for consumer television receiving installations.

1.3.2 Polarization of electromagnetic waves

Electromagnetic waves radiated by a transmitter are characterized by their frequency and polarization. Polarization defines the orientation of the electric and magnetic field components (which are orthogonal to each other and to the direction of propagation) of the electromagnetic field. Polarization is determined by the characteristics of the radiating device(s) of the transmission antenna.

Two different kinds of polarization are used for satellite transmissions.

(1) *Linear polarization*
This is characterized by a fixed orientation of the electric and magnetic field vectors independently of the position on the propagation axis (see Fig. 1.10).

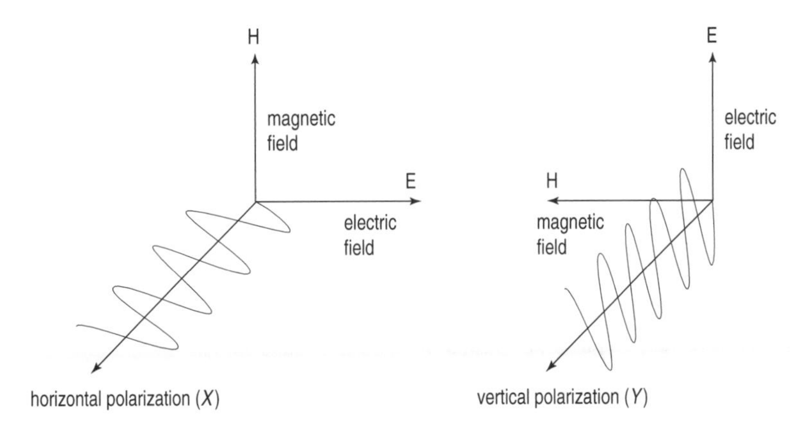

Fig. 1.10 Illustration of linear polarization

The polarization of the wave is named after the orientation of the electric field vector: if it is 'vertical', then we say that the wave is vertically polarized, if it is 'horizontal', then we say that the wave is horizontally polarized (see Note 1.1).

It is thus, in principle, possible to transmit two waves of identical frequency and of orthogonal polarizations (let us say vertical and horizontal), modulated by different signals, with no interference between them, since it will be possible to separate the waves at the receiving end by means of an *accurate* alignment of their orientation around the propagation axis of the wave collector (the LNB).

This allows us practically to double the transmission capacity in a given frequency band.

Note 1.1

The two orthogonal polarizations (normally named *X* and *Y*) are only seen as 'vertical' and 'horizontal' by a station located on the same meridian as the satellite, unless a deliberate polarization offset has been applied in order to obtain vertical and horizontal polarizations at the centre of the target area. It is not located on the same meridian as the satellite, as is often the case.

(2) *Circular polarization*

In this case, the orientation of the two field vectors is not fixed along the propagation axis, but rotates along this axis at the rate of one turn (360°) for each wavelength (see Fig. 1.11). However,

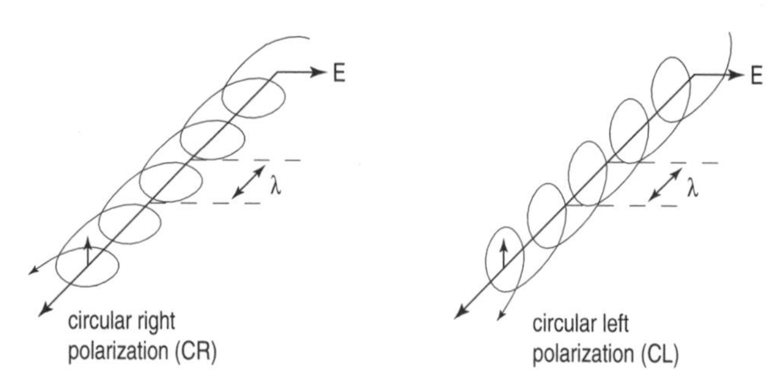

Fig. 1.11 Illustration of circular polarization

the two vectors are still orthogonal to each other at any place on this axis.

Depending on the sense of the rotation around the axis, the polarization is named 'circular left' or 'circular right', and the polarizations can be separated in the same way as for linear polarization, by the wave collecting device, thus also allowing for transmission capacity doubling.

The advantage of circular polarization compared with linear polarization is that it does not need an accurate positioning of the wave collecting device, and the separation between circular left and circular right polarizations is determined only by the accuracy of the collecting device.

A slight disadvantage is that, in the Ku band, circular polarization suffers from a more noticeable depolarization in adverse weather conditions, which, in this case, reduces the separation between the two polarizations (cross-polarization) and thus can degrade the receiving signal more than in the case of linear polarization. This is one of the reasons (although not the main one) why circular polarization is tending to disappear in the Ku band for the new (mainly digital) satellite transmissions in Europe.

1.4 Transmission: basic principles

In order to understand the main concepts used for calculation of a reception installation, it is important to know some relatively simple principles that rule the transmission of energy radiated by an electromagnetic wave transmitter in free space (as in the case of broadcasting satellites, which are located in a vacuum and at a relatively large distance from any other object).

Let us assume a transmitter with a radiated power P, located in free space at a distance R from the place of reception, and radiat-

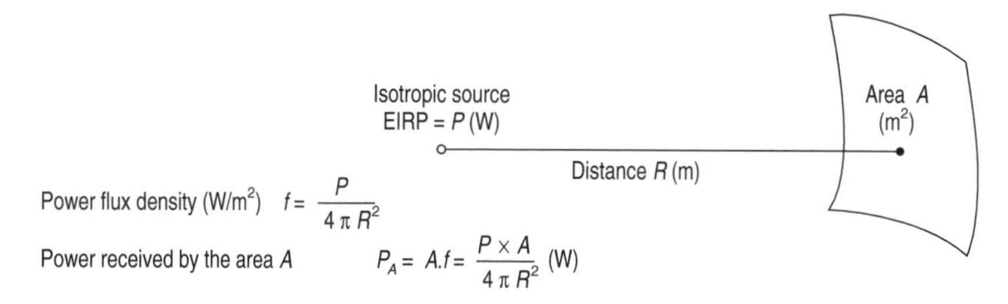

Fig. 1.12 Power flux of an isotropic source

ing uniformly in all directions (such an ideal transmitter is said to be an *isotropic* source) (Fig. 1.12).

Since there is no loss of energy in the vacuum, at a distance R from the source, the energy will be uniformly distributed on a sphere of radius R, of which the area S is equal to $4\pi R^2$.

If P is expressed in watts and R in metres, the *flux density* f through any area A, expressed in W/m² will then be:

$$f = P/S = P/4\pi\ R^2\ [\mathrm{W/m^2}] \tag{1.1}$$

So, the total power received by the area A will be:

$$P_A = fA = PA/4\pi\ R^2\ [\mathrm{W}]$$

In practice, a true isotropic source is not realizable and, in this case, not even desired. Since the electric power is very limited on board the satellite, it is of utmost importance not to disperse it in all directions, but to concentrate it as precisely as possible on the 'target' service area by means of a directive transmission antenna, in order that the electromagnetic field be maximal and homogeneous in this area and as small as possible (ideally zero) outside the service area.

1.4.1 Equivalent Isotropic Radiated Power (EIRP)

Reminder

It is the general rule to express in decibels (dB) the ratios between RF electrical powers. This allows, among other things, to simplify gain or loss calculations by replacing multiplications or divisions by additions or subtractions.

The power ratio (or relative power) $G_{2/1}$ expressed in dB of two powers P_2 and P_1 is given by:

$$G_{2/1}\ [\text{dB}] = 10 \log (P_2/P_1)$$

In order to indicate the intensity of the satellite signal in a given place (which it is necessary to know in order to determine the receiving antenna characteristics), it is common practice to publish 'level curves' characterized by a figure called the **EIRP** (equivalent isotropic radiated power). EIRP is expressed in dBW (decibel. watt), which means 'decibel relative to one watt'.

This concept is relatively simple.

The EIRP (in dBW) of a satellite at a given reception point is the power relative to 1 W that would be required from an isotropic source situated at the same place as the satellite to produce the same flux density as that received from the satellite, at the same reception point.

This figure is made up of two components: it is the sum of the gain, G_E, in radiated electric power (between the satellite's transmitter and a 1 W transmitter) and the transmitting antenna gain, G_A, due to the directivity (relative to the isotropic antenna).

$$\text{EIRP} = G_E + G_A \qquad (1.2)$$

Calculation of EIRP in a simplified case

Let us assume an ideal satellite radiating a power of 100 W uniformly on a circular area, 1000 km radius, on the Earth, and nothing outside this area (Fig. 1.13).

In order to simplify calculations, we will assume that this area is centred on the equator, just 'below' the satellite; this area is located at approximately 36 000 km from the satellite, is orthogonal to the radiation direction of the transmitter, and it can be considered as being a flat area.

The electrical power gain of the 100 W transmitter G_E (in dB) relative to 1 W (P_1) is:

$$G_E = 10 \log (P_E/P_1) = 10 \log(100) = 20 \text{ dB}$$

An ideal isotropic transmitter would radiate uniformly its whole power in a spherical area which, at the level of the Earth's surface would have a radius R = 36 000 km.

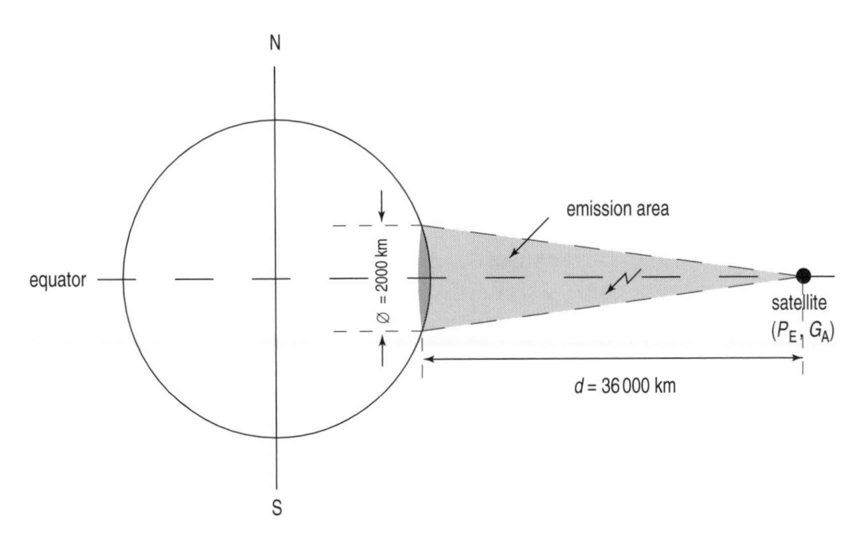

Fig. 1.13 Illustration of EIRP calculation

The area S of this sphere, which uniformly receives all the radiated energy is $4\pi R^2$, so:

$$S = 4\pi\ (36\ 000)^2\ \text{km}^2$$

Our hypothetical directional satellite radiates all its power in a cone which, at the level of the Earth 'cuts' a circular area of radius r_Z = 1000 km, which we will consider as being flat.

The area of this circular zone in which all the energy is concentrated is $s_Z = \pi\ r_Z^{\ 2}$, so:

$$s_Z = \pi\ (1000)^2\ \text{km}^2$$

The 'gain' G_A brought by the directivity of the transmitting antenna represents the power ratio that would be necessary to obtain the same power flux density with an isotropic transmitting antenna. It is thus inversely proportional to the illuminated areas, which in dB gives:

$$G_A = 10 \log (S/s_Z) = 10 \log [4\pi\ (36\ 000)^2/\pi\ (1000)^2]$$
$$= 10 \log [4 \times (36)^2] = 37.1\ \text{dB}$$

As indicated before, the EIRP is the sum of the electric power gain, G_E, of the transmitter relative to 1 W and the gain, G_A, of the transmission antenna. Consequently, the resulting power relative to 1 W, which is expressed in decibel.watt (dBW) is, in this case

$$\text{EIRP} = G_E + G_A = 20 + 37.1 = 57.1 \text{ dBW}$$

1.4.2 Received power level

Let us continue our example, by assuming, always in order to simplify calculations, a reception with a (parabolic) antenna of diameter 1 m and with an efficiency of 100% (which means that all the energy received by the reflector will be transmitted to the receiver). We will also assume that transmission between the satellite and the antenna does not involve any energy loss (which is true for the major part of the trajectory, which takes place in a vacuum, but not completely for the atmospheric part, especially in the case of precipitation).

The power received by the antenna will be equal to the product of the transmission power and the ratio of the apparent area of the antenna (of radius r_a = 0.5 m) and of the area s_z of the 'illuminated' zone (of radius r_z = 1000 km = 10^6 m), since these two zones are both orthogonal to the radiation direction because we are located at the equator, just below the satellite (Fig. 1.14).

(Another method would be to calculate the flux density and multiply by the area of the antenna.)

In our example, the apparent antenna area is:

$$s_a = \pi r_a^2 = \pi (0.5)^2 = 0.25 \pi \text{ [m}^2\text{]}$$

and the area of the illuminated zone is:

$$s_z = \pi r_z^2 = \pi (10^6)^2 = 10^{12} \pi \text{ [m}^2\text{]}$$

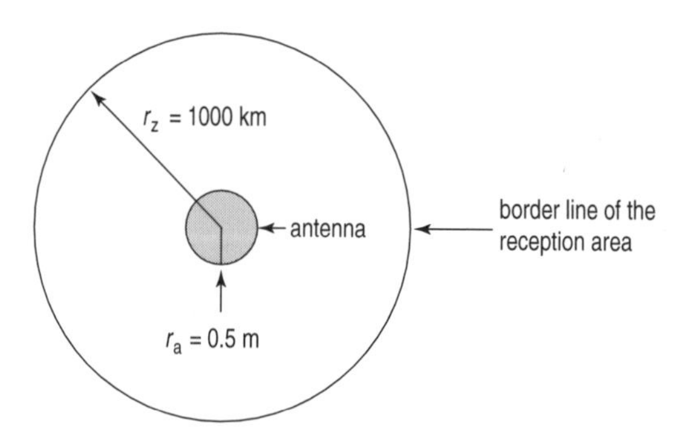

Fig. 1.14 Relative areas of the antenna and the reception area (seen from the satellite)

The ratio between the power, P_R, of the signal received by the antenna and the transmitted power, P_E, uniformly spread over the illuminated area is equal to the ratio of the antenna area s_a and the area of the illuminated zone s_z:

$$P_R/P_E = s_a/s_z = 0.25\pi/10^{12}\pi = 0.25 \times 10^{-12}$$

From this, we can calculate the 'directional' attenuation A_d in dB (which is a negative gain)

$$A_d = -10 \log P_R/P_E = -10 \log (0.25 \times 10^{-12}) = 126 \text{ dB}$$

In this idealized case, with our 100 W transmitter ($p_E = 10 \log P_E = 20$ dBW), the power of the received signal (in dBW) would be $p_R = 10 \log (P_R)$ or

$$p_R = p_E - A_d = +20 - 126 = -106 \text{ dBW}$$

This figure gives an idea of the very low value of the received signal (in fact it is even weaker in reality since we neglected all losses): -106 dBW means '106 dB below 1 W', which corresponds to a received power of 0.25×10^{-10} W!

1.4.3 Isotropic transmission attenuation

In order to simplify antenna calculations and to be independent of the radiation characteristics of the transmitting and receiving antennas, it is convenient to define an isotropic transmission attenuation, A_i (which assumes isotropic transmitting and receiving antennas, i.e. by definition, without any gain), which is valid for a satellite in geostationary orbit.

It can be demonstrated that this attenuation obeys the following formula:

$$A_i = 20 \log (4\pi R/\lambda) \tag{1.3}$$

For a frequency of 4 GHz (near the middle of the C band), we have $\lambda = 7.5 \times 10^{-2}$ m.

At the equatorial point just below the satellite, $R \approx 36\,000$ km $= 36 \times 10^6$ m, then

$$A_i = 20 \log (4\pi \times 36 \times 10^6/7.5 \times 10^{-2}) \approx 195 \text{ dB}$$

At 12 GHz (near the middle of the Ku band), we have $\lambda = 2.5 \times 10^{-2}$ m.

In the same conditions as above, this gives

$$A_i = 20 \log (4\pi \times 36 \times 10^6/2.5 \times 10^{-2}) \approx 205 \text{ dB}$$

Exercise—How to calculate this result from the example of the previous paragraphs? We have calculated a (directive) attenuation of 126 dB in the previous case (a directional geostationary satellite of 100 W radiating a wave at 12 GHz onto an area with a radius of 1000 km centred below the satellite and received with an ideal antenna of diameter 1 m).

This reception antenna is also directional and exhibits a gain relative to the isotropic antenna. It can be demonstrated that, for a wavelength λ, which is relatively small compared with the antenna diameter, the theoretical gain (i.e. with an efficiency of 100%) relative to the isotropic antenna of an antenna of area s_a, is given by the relation

$$G_R = 10 \log (4\pi s_a/\lambda^2) \tag{1.4}$$

For a frequency of 12 GHz, $\lambda = 2.5 \times 10^{-2}$ m, so $\lambda^2 = 6.25 \times 10^{-4}$ m^2; the apparent area s_a of the antenna (of diameter 1 m) is $s_a = 0.25\ \pi$ m^2, so:

$$G_R = 10 \log (4\pi\ s_a/\lambda^2) = 10 \log (10^4\ \pi^2/6.25)$$
$$= 10 \log (15\ 776) \approx 41.9 \text{ dB}$$

If we want to consider the case of an isotropic transmission source and an isotropic reception antenna, then we have to add to the directive attenuation calculated before (A_d = 126 dB) the gain of the transmission antenna calculated previously (G_A = 37.1 dB) and the gain of the receiving antenna (G_R = 41.9 dB), to find the isotropic attenuation A_i

$$A_i = A_d + G_A + G_R = 126 + 37.1 + 41.9 = 205 \text{ dB}$$

This result confirms the result of the theoretical formula.

In Europe, the real figure has to be increased by approximately 1 dB since the distance from the satellite is greater (approximately 39 000 km): the isotropic attenuation (atmospheric attenuation excluded) is then approximately 206 dB.

Isotropic attenuation will then be used for calculation of the power of the received signal p_R from the EIRP and the gain G_R of the reception installation by means of the relation:

$$p_R \text{ [dBW]} = \text{EIRP} - A_i + G_R \qquad (1.5)$$

In a real case, all 'secondary' attenuations neglected up to now will have to be taken into account, as will be explained in section 3.4.

1.4.4 Uplink

The *uplink*, as its name implies, is the link between the terrestrial transmission and control station and the satellite; the satellite contains a certain number of *transponders* the function of which is to retransmit the TV signals to a certain service area of the Earth.

The uplink is modulated in the same way and occupies the same bandwidth as the *downlink*, each transponder being 'transparent' to the signal. Its main task is to make a simple frequency change followed by a power amplification of that part of the frequency spectrum that corresponds to the channel to be retransmitted to the service area (Fig. 1.15).

Frequencies used by the uplinks of satellites operating in the Ku bands are generally within the bands from 14 000 to 14 800 GHz and from 17 300 to 18 100 GHz. The uplink must obviously not perceptibly degrade the quality of the TV signal. For this purpose, a very 'comfortable' signal-to-noise ratio must be ensured at the input of the transponder's receiver, which, however, cannot have an antenna of much more than 3 m diameter. That is why the power of the terrestrial station is important (of the order of 1 kW, so 30 dBW) and is combined with a large gain antenna (> 60 dB), which has also to be very directional in order not to interfere with neighbouring satellites. This implies large dimensions (> 10 m). This corresponds, for the terrestrial station, to an EIRP of around 90 dBW, which is 30 to 40 dBW more than the EIRP of direct broadcast satellites towards the Earth!

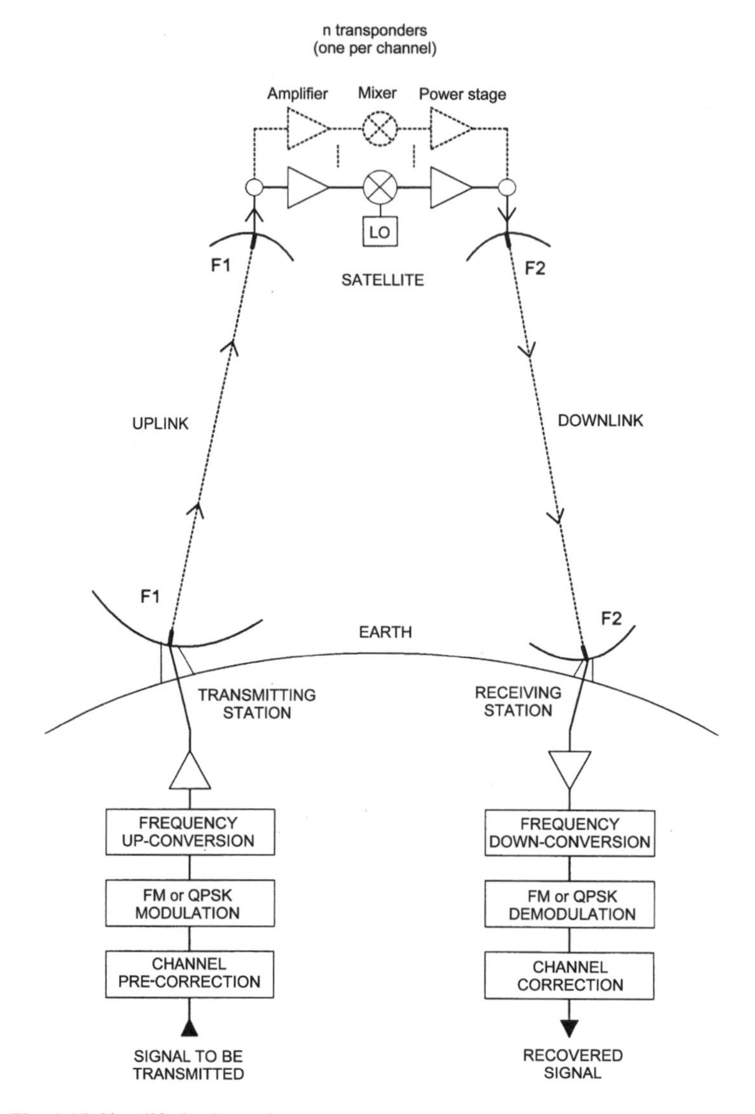

Fig. 1.15 Simplified schematic representation of a satellite TV transmission system

2 Direct-to-home satellites

Geostationary satellites were initially used mainly for intercontinental telecommunications (telephone, data and fixed or occasional links beween radio and television broadcasting organizations). Then, from the beginning of the 1970s, they were used in the United States by cable television companies to supply TV programmes to their various 'head end' stations scattered all over the country.

Since, for many years, these transmissions were not scrambled, this gave birth to a kind of 'pirate' direct reception by more and more amateurs. This is why large antennas (3 to 4 m in diameter) have flourished in many US backyards—the EIRP of these satellites (of the order of 40 dBW) operating in the C band (3.7 to 4.2 GHz) was not intended for direct reception.

The Ku band was then used in most countries for fixed telecommunication links (the so-called FSS (Fixed Satellite Service) bands: 10.950 to 11.700 and 12.500 to 12.750 GHz) with, among others, occasional or permanent TV 'feeds' that have attracted the attention of amateurs interested in **DXTV** reception and not-so-ordinary transmissions.

2.1 The high power satellite epoch

The idea to launch satellites for direct diffusion of TV programmes to 'ordinary' people (a service named **DTH** for 'Direct-To-Home') appeared economically realizable at the beginning of the 1970s. The technical specifications of such a service were defined (except for the USA) by the World Administration

Radiocommunication Conference which met in 1977 (**WARC** 77) which also defined the frequency bands, transponder power and polarization to be used, and allocated orbital positions, channels and EIRP contours to each of the participating countries.

The attributed frequency band (from 11.700 to 12.500 GHz for Europe, Asia and Africa) is known as the **DBS** band (Direct Broadcast Satellite) or **BSS** band (Broadcast Satellite Service) in the USA. It has been chosen because, on the one hand, it allows well delimited service areas with relatively small transmission antennas on the satellite (less than 3 m diameter) and, on the other hand, because it does not suffer too much from meteorological conditions. Although it is only 800 MHz wide, the DBS band has been split into 40 channels of 27 MHz width placed on a grid of 19.18 MHz (see Fig. 2.1 and Table A.0 in the Appendix).

In fact, as explained in the previous chapter, it is possible to use the same frequency to transmit two different programmes on two orthogonal polarizations, with the condition that the receiving antenna ensures a separation of at least 20 dB between them. However, for a consumer service, it has been preferred to have a bigger protection margin between adjacent channels by providing a 'guard space' between consecutive channels of the same polarization and by interleaving the channels of opposed polarization on this guard space (see Fig. 2.1). In this way it has been possible to place 40 channels within 800 MHz, which otherwise would have occupied approximately 1200 MHz (40 × 27 = 1080 MHz + 10% for guard spaces).

Circular polarization (right for odd channels, left for even channels) was preferred over the linear polarization used by telecommunication satellites. This is because, for circular polarization, the orientation of the reception head (LNB) around the propagation axis is unimportant, and therefore does not require

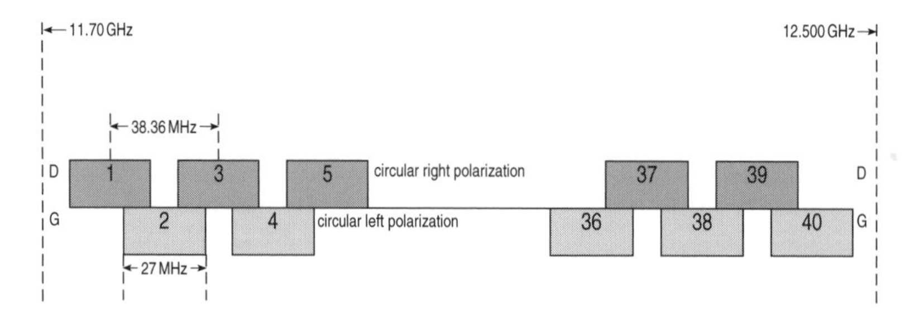

Fig. 2.1 WARC 77 DBS channel configuration

any precise adjustment; this point is especially important for motorized antennas which, with linear polarization, require a polarization adjustment (*skew*) for each different satellite. Eight orbital positions, with 6° between them (from 9°E to 37°W) have been defined for Europe; however, only two of them (19°W and 31°W) have really been used, and even then only partly. Eight countries were planned to 'cohabit' on each position, with five channels allocated to each of them. The angular distance between adjacent satellites operating in this band was chosen to be large enough (minimum 6°) to avoid any interference, even with antennas that are not very directional (instead of the 3° generally separating two adjacent telecommunication satellites).

The orbital position attribution has been chosen in such a way that the possible transmission interruption due to the solar equinox eclipse happens very early in the morning for the targeted service area (between 2 and 4 a.m.) to minimize disturbance.

The channel allocation (odd or even) has also been done in such a way that countries speaking the same language were kept, as much as possible, on the same polarization of the same satellite, in order to simplify the installation (and also to avoid the need for a mechanical polarizer, which was rather costly at that time). However, this has not been possible for those countries that speak more than one language (Switzerland and Belgium among others). Last, but not least, the power of the satellites was defined to be high enough to allow reception by means of antennas of less than 90 cm in the main service area. Taking into account the poor performances of the LNBs at this time, this led to a very high power for the transponders (more than 200 W–230 W for the 'twin' French and German satellites TDF and TV-SAT).

As can be seen, everything seemed very carefully worked out and guaranteed to ensure success. Actually, at that time (remember: 1977), telecommunication deregulation was not the universal motto and, in addition, for some countries, one of the objectives of these satellites was to fill gaps in the coverage of the terrestrial transmitter network (which in most cases carried only two or three programmes).

However, the construction of these high power satellites suffered significant delays compared with the original plans, and even after that they encountered many problems. TV-SAT1 launched in November 1987 on 19°W did not work due to the bad opening of one of its solar panels; TV-SAT2 was launched in August 1989 and worked until 1994 with three D2MAC 'clear' channels and one channel with 16 digital radio programmes using the now defunct **DSR** German norm (Digital Satellite Radio). At the same position of 19°W, TDF1 was launched in October 1988 and TDF2 in July 1990; they operated with a number of D2MAC channels, which changed from four at the beginning of its life to one at the end (mid 1997) because of reliability problems with the power stage's Travelling Wave Tubes (**TWT**). Marco-Polo, launched at the same time on 31°W, very quickly ceased its D-MAC transmission due to the bankruptcy of its UK operator, BSB, which merged with Sky TV (its main competitor on ASTRA using PAL/Videocrypt) to form the current analogue BSkyB company.

Olympus 1, an experimental satellite launched in July 1989 on 19°W, definitely ceased transmissions in 1992 after having been out of control for two months and then 'saved' in July 1991. Only Hispasat 1 and 2, launched in 1992 and 1993 on 31°W and less powerful (110 W), still work as expected and are partly used for analogue PAL and more and more for MPEG-2/DVB transmissions. Some Scandinavian satellites from this type are also still in operation in D2MAC. All other European countries never used their DBS channel allocation on these orbital positions.

To all these technical problems, a number of political or economic problems have been added:

- D2MAC and DMAC, the new standards straight out of the laboratories in the mid-1980s, were imposed by Brussels for DTH transmissions in the DBS band; this has delayed the availability and increased the price of the receivers, in view of a hypothetical compatibility with a future HDTV standard (HDMAC);
- cable TV started its deployment over Europe, and a number of new private terrestrial channels were launched in most European countries.

This finally led to the complete failure of this type of satellite, of their orbital positions as well as the D2MAC standard, and also to the main profit of ASTRA which, at the same time, started transmissions in PAL in the FSS band on 19.2°E. TV-SAT2 was repositioned in 1995 on 1°W for a Scandinavian operator, while TDF2, after having been used with only a transponder for a long time, was sold to Eutelsat (mid 1997) and repositioned on 36°E for a Russian operator, where it is supposed to ensure two years of service before the launch of a new satellite.

2.2 Television 'squats' on the telecommunication bands

The specifications of direct-to-home television satellites have evolved considerably since the WARC 77 plan specifications were defined:

- on one hand, the reception heads (LNB) have made enormous progress, with average noise factors in the range of 1.5 to 1.8 dB in 1990 and 0.8 to 1.1 dB in 1998, instead of 3 to 5 dB in 1977, which ensures the same service area with a much smaller transmission power (50 to 100 W instead of 200 W) at identical or smaller antenna dimensions.
- on the other hand, the increase in useful load of the launching rockets has allowed an important increase in the total electrical power available, which, coupled with the transponder power reduction, has allowed the number of transponders per satellite to be increased by a factor of 4 (up to more than 20 transponders of 100 W each in the last satellites or 'birds'). These satellites now have a sufficient electrical autonomy to ensure an uninterrupted and normal service during equinox solar eclipses.
- the lifetime of the satellites has doubled, from approximately 7 to 15 years, partly due to the better accuracy of the launches, which strongly influences the quantity of remaining propellent necessary for position control of the satellite.

Certain of the failure of the high power satellites of the WARC 77 concept (i.e. designed on the basis of outdated technological and political data), the Luxembourg based SES (Société Européenne de Satellites) decided, somewhat in violation of the international conventions of the time, to launch medium power (50 W) direct-to-home television satellites operating in the FSS band, on the 19.2°E orbital position.

The first of these, ASTRA 1A, was launched in December 1988 and had sixteen 50 W transponders, ensuring an EIRP of 52 dBW over an area of more than 1000 km radius, which made reception in this region possible with a 60 cm dish (with an LNB of less than 2 dB noise factor). The channel width was fixed at 26 MHz on a grid of 14.75 MHz, in linear polarization (horizontal/vertical) (see Fig. 2.2).

Success was almost immediate, first in the United Kingdom with the pay-TV channels of BSkyB, in Germany with free-to-air programmes, and some Scandinavian channels in D2MAC.

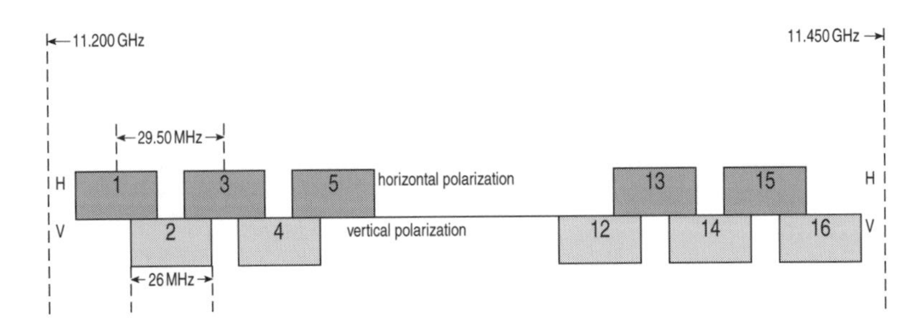

Fig. 2.2 Configuration of the ASTRA 1A channels (1B, 1C, 1D occupy three identical contiguous bands)

ASTRA 1B and 1C followed in 1991 and 1993 respectively, each adding 16 new channels up to the complete occupancy of the FSS band (10.950 to 11.700 GHz) with 48 channels. ASTRA 1D was then launched in 1994, bringing the channel number to 64 on this orbital position and extending the FSS band down to 10.700 GHz.

The Scandinavian D2MAC transmissions have now moved to other satellites, so the TV standard used by analogue transmissions on ASTRA is now exclusively PAL. Table A.1 (Appendices) details the 64 channels in the FSS extended band of the ASTRA system.

The Eutelsat satellite system on 13°E (Eutelsat II-F1 and the 'Hot Bird' series) which, with ASTRA, is the second most important for Europe, has chosen a relatively similar approach, although its missions are more diverse.

The currently used FSS band now goes down to 10.700 GHz, due to an extension of thirteen 33 MHz channels (110 to 122) with Hot Bird 5.

The Eutelsat satellites use different types of beams depending on their mission and generation (Eutelsat II-F1 or the Hot Bird series), and a new channel numbering system has been developed ('normal' channels of 33 MHz on a grid of 19.18 MHz and 36 MHz on a grid of 20.75 MHz, and some 'wide' channels of 50 or 72 MHz, see Table A.2).

The French Telecom 2A (8°W), Telecom 2B (5°W) and Telecom 2C (3°E) satellites, launched in 1991 and 1992 use the high portion of the FSS band, often therefore referred to as the 'Telecom' band (12.5–12.75 GHz); this band has been divided into 11 channels of 36 MHz on a grid of 21 MHz (Fig. 2.3). There, the five main free-to-air French SECAM channels can be received (Telecom 2B) along with the analogue pay-TV Canal Satellite collection or 'bouquet' (Telecom 2A, although probably not beyond the end

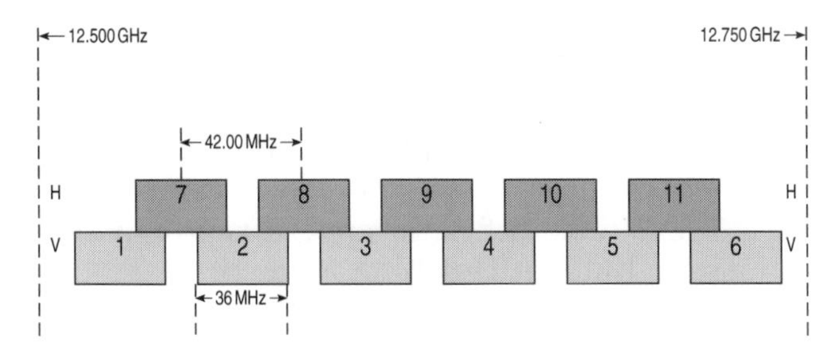

Fig. 2.3 Configuration of the Telecom 2A/2B/2C channels

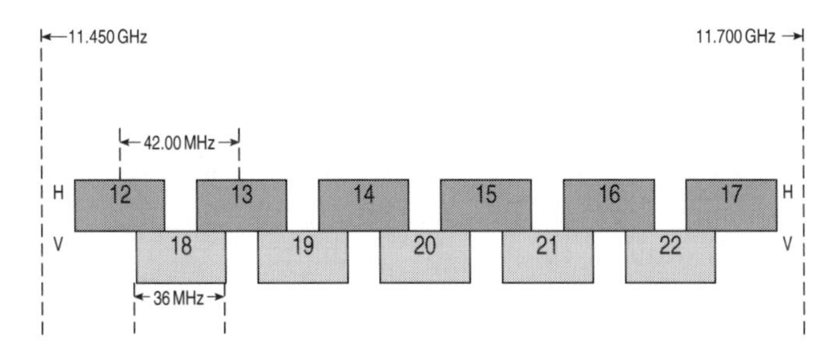

Fig. 2.4 Configuration of the Telecom 2D channels

of the century). Telecom 2C (on 3°E) broadcasts mainly scrambled programmes for cable head-ends and 'feeds'.

The Telecom 2D satellite (launched in August 1996) added 11 new identical channels (36 MHz) in the low FSS band (see Fig. 2.4). At the time of writing these channels were used mainly for 'feeds' (analogue or digital). It seems that the direct broadcast missions of the Telecom satellites will be more and more restricted, especially as all new French digital channels use either ASTRA or Eutelsat systems.

Table A.3 details the 22 channels in the Ku band of the Telecom system.

2.3 Digital television: the return of the DBS band

The start of digital television transmissions in 1996 on ASTRA and Eutelsat gave new life to the DBS band, which has been divided into 40 channels of 33 MHz in linear polarization, in order to allow—in a simple way—reception with the same LNB as that used for the FSS band (instead of 27 MHz and circular polarization, as defined by WARC 77).

The DBS channels of ASTRA and Eutelsat are, however, not exactly the same: Eutelsat has kept the 19.18 MHz channel step of the WARC 77 plane while ASTRA chose a step of 19.5 MHz. Figure 2.5, and Table A.4 in the appendix, give frequencies and polarizations for ASTRA 1E and 1F. ASTRA 2A and 2B on 28.2°E, which carry among others the BSkyB digital bouquet, use the same frequency plan.

Figure 2.6, and Table A.5 in the appendix, give frequencies and polarizations for Eutelsat Hot Birds 2 and 3.

In addition, ASTRA 1G (Fig. 2.7) exploits the high FSS band (12.5–12.75 GHz) with 16 channels of 26 MHz, similar to those used by ASTRA 1A to 1D, but with an asymmetrical offset between opposed polarizations

Eutelsat's Hot Birds 4 and 5 (Fig. 2.8) divide this same band, 12.5–12.75 GHz, into 12 channels of 33 MHz, identical to those used in the DBS band (see Note 1).

Unlike ASTRA, on Eutelsat Hot Birds there is currently no fixed frequency frontier between analogue and digital services, which coexist almost all over the Ku band. It is, however, certain that analogue transmissions will be progressively replaced by digital services, which sooner or later will occupy the complete Ku band on all satellites.

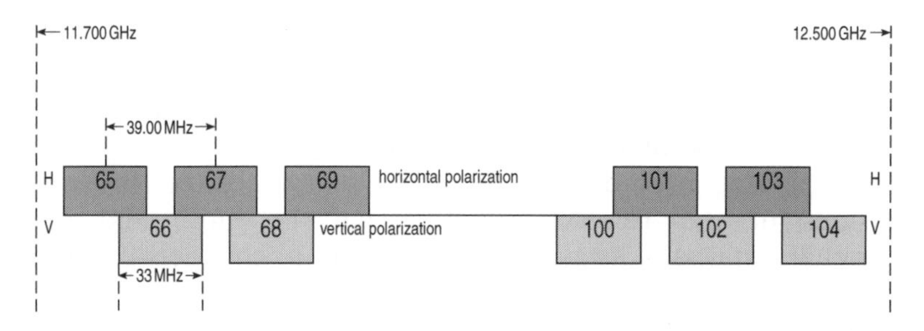

Fig. 2.5 Configuration of the ASTRA (1E/1F and 2A/2B) digital channels in the DBS band

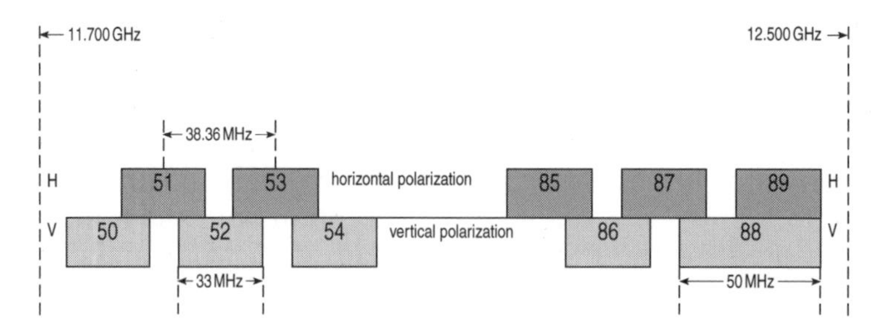

Fig. 2.6 Configuration of the Eutelsat (Hot Birds 2 and 3) channels in the DBS band

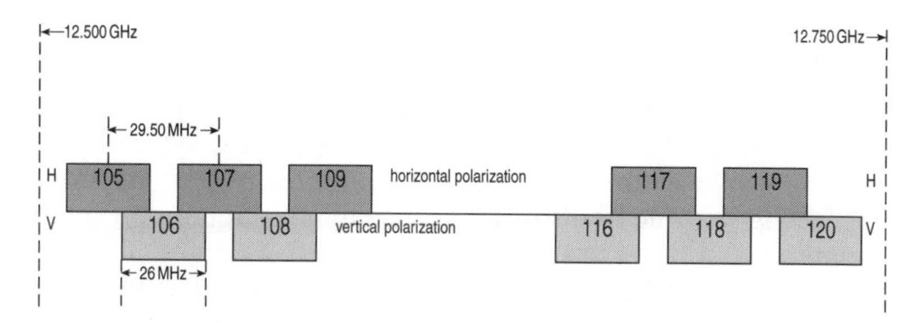

Fig. 2.7 Configuration of the ASTRA 1G/2B digital channels in the high FSS band

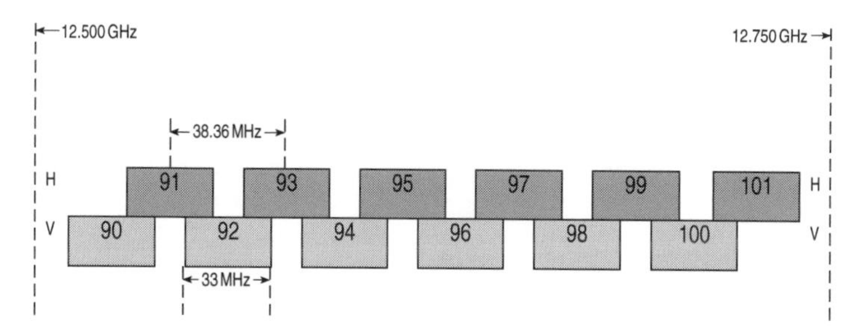

Fig. 2.8 Configuration of the Eutelsat channels (Hot Birds 4 and 5) in the high FSS band

2.4 Service areas

In the previous chapter, we studied a simple ideal case where the service area was circular, with abrupt limits to that area. In practice, things are somewhat more complex: on one hand, today it is possible to define service areas that have a relatively complex shape, i.e. following rather accurately the coasts and borders of a country or a group of target countries, and on the other hand, there is always a transition area where the received signal decreases progressively, although sometimes rather quickly. In these real cases, the concept of EIRP is very helpful, since the iso-EIRP curves published by the satellite operators quickly give a good idea of the receiving conditions in a given place. Some examples are given in Appendix A.1 for ASTRA 1 and 2 series and Eutelsat 'Hot Birds'. Some operators, such as ASTRA, publish curves directly indicating the antenna diameter rather than the EIRP, which has the advantage of saving calculations; however, it has the disadvantage that the exact conditions (LNB noise figures, various losses, security margins) are not given explicitly.

Note 1

In digital television with DVB parameters, a 33 MHz channel allows a *symbol rate* of 27.5 Msymb/s (million symbols per second), while a 26 MHz channel is limited to 22 Msymb/s. Depending on the multiplexing system and the programme content, this allows transmission of seven to eight TV programmes of good quality in the first case and five to six in the second case if the most frequent *code rate* is used ($R_c = 3/4$).

3 Capture of the satellite signal

3.1 Antenna

We saw earlier that the signals received from the satellites, which are more than 36 000 km away from the Earth, are extremely weak; thus, the antenna is a critical element in the reception chain, and has to be considered very seriously in order to obtain satisfactory results.

The parabolic antenna, in its many forms, is by far the most popular antenna for satellite TV reception. This is due to the remarkable property of the parabolic mirror of concentrating onto a point (called the *focus*) an incident beam (be it light or other kinds of waves) parallel to the mirror's axis. Other, much less used, antenna shapes exist, which we will mention only for completeness.

3.1.1 The parabolic antenna

The parabolic antenna is made of a reflecting mirror that concentrates the incident waves onto its focus, where the low noise block converter, **LNB** or **LNC** (which is the active part of the antenna), is located. The LNB consists of a selective amplifier and a frequency converter, which lowers the incoming frequency range (of the order of 12 GHz, a frequency too high for transmission by a coaxial cable) to a frequency range called the Satellite Intermediate Frequency (**SAT-IF**), which is in the range of 1 to 2 GHz, and which will be applied to the receiver's input.

The mirror is a 'paraboloid of revolution', which is obtained by rotating a parabola around its axis (of which the general form of the equation is $y = kx^2$ when it is centred on the y axis).

It can be demonstrated that a parabolic mirror reflects and concentrates the incident rays parallel to its axis on a point called the focus (see Figs 3.1 and 3.2), situated on the y axis at a distance f from the parabola peak, for example $f = 1/4k$. The equation of the parabola can thus be written $y = x^2/4f$.

The distance (measured from a plane perpendicular to the axis) covered by all the rays of a beam parallel to the axis and reflected

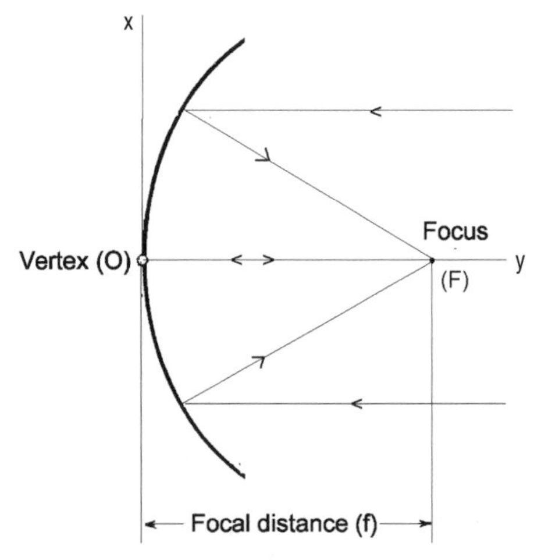

Fig. 3.1 Concentration of a parallel beam to the focus of a parabola

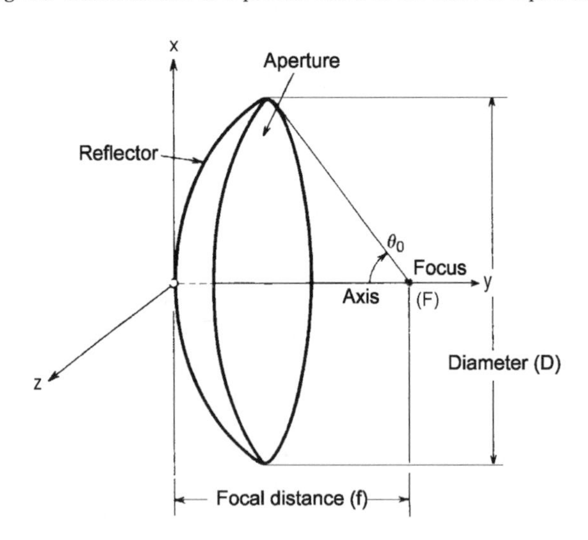

Fig. 3.2 Main characteristics of a parabolic mirror

by the mirror is the same when the rays reach the focus. Thus, if the rays are electromagnetic waves, they all will arrive in phase at the focus, and their energies will be combined.

One can demonstrate that, for a relatively small wavelength, λ, compared with the dimensions of the antenna, the theoretical gain (i.e. with an efficiency of 100%) of a parabolic antenna of surface s_a relative to the isotropic antenna is given by the relation $G = 4\pi s_a/\lambda^2$, thus in dB

$$G_R = 10 \log (4\pi s_a/\lambda^2)$$

The angular directivity (at −3 dB) of a parabolic antenna illuminated by a circular source is approximately:

$$d\,(°) = 72\,\lambda/\Phi$$

(where Φ is the diameter in cm).

Figure 3.3 represents an example of a directivity diagram (the gain of the antenna as a function of the angle relative to the axis). In addition to the main directivity lobe, secondary lobes can be observed. These are due to various parasitic effects (unevenness of the mirror surface, edge effects etc).

Table 3.1 indicates the theoretical gain and directivity at 12 GHz, calculated from the above relations, for antenna

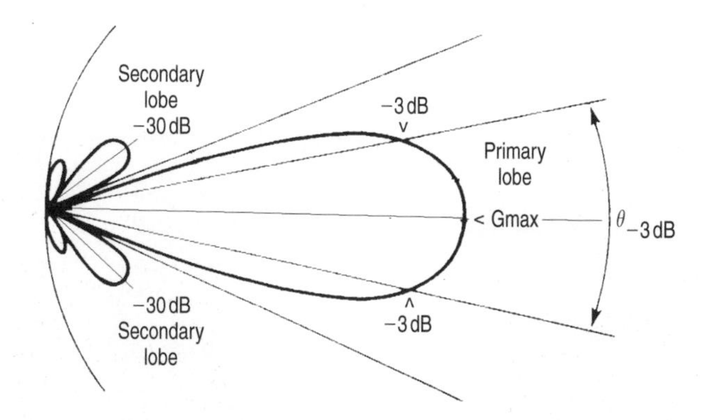

Fig. 3.3 Directivity diagram of a parabolic antenna

Table 3.1 Theoretical gain and directivity of a parabolic antenna at 12 GHz

Φ (m)	0.4	0.5	0.6	0.7	0.8	0.9	1.0
Gain (dB)	34.0	35.9	37.5	38.9	40.0	41.1	41.9
Dir. −3 dB (°)	4.5	3.6	3.0	2.6	2.2	2.0	1.8

diameters from 40 cm to 1 m (the maximum value not requiring special permission). For antennas of greater dimensions, the gain can be found by taking the value for an antenna half its diameter and adding 6 dB, and the directivity can be found by halving its angle.

Practical antennas have an efficiency between 55 and 85%, which Table 3.2 translates into a loss (in dB), to be subtracted from the theoretical gain mentioned above.

Table 3.2 Loss (dB) as a function of the antenna efficiency

η (%)	55	60	65	70	75	80	85
Loss (dB)	2.59	2.21	1.87	1.55	1.25	0.96	0.70

3.1.2 Various types of parabolic antennas

Four main types of parabolic antennas have been developed, which differ mainly in their complexity and efficiency.

1. The '*prime focus*' antenna: this is the direct application of the parabola theory (Fig 3.2), and thus historically the first used for individual reception (see Fig. 3.4).

   The LNB is placed at the main focus point by means of a 'tripod' attached to the periphery of the mirror, and the axis of the parabola has to be directed towards the satellite. Consequently, the LNB and the tripod make a 'shadow' on the mirror, thus reducing the effective area of the mirror and the

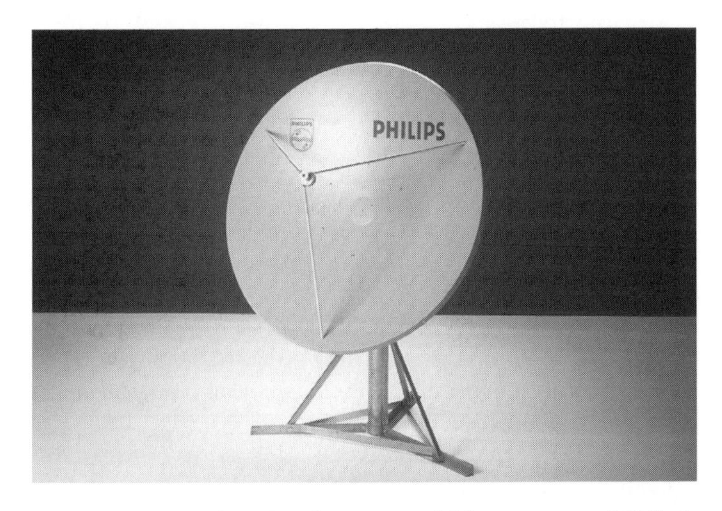

Fig. 3.4 Example of a 'prime focus antenna' (photo courtesy of Philips)

antenna efficiency. This effect is of greater importance for small antennas, which is why this type of antenna is seldomly used in sizes smaller than 1 m diameter.

Depending on the quality range, the mirror can be made of a simple galvanized iron sheet, an aluminium sheet or a composite covered by a conducting layer. The efficiency is generally between 55 and 65% depending on the dimensions and implementation.

It should be noted that the angle of the mirror of this type of antenna to the vertical is equal to the elevation of the satellite, of the order of 25 to 40° in most European latitudes. It can consequently accumulate snow in winter, which can reduce the gain of the antenna up to the point where reception is interrupted. That is why a variant with an offset focus is now often preferred.

2. The *offset focus* antenna (often simply called 'offset') in fact uses a reflector cut from prime focus parabolic mirror of larger dimensions, in such a way that the LNB is situated outside the path of the incoming beam to the mirror, thus eliminating the shadow of the LNB on the mirror (see Figs 3.5(a) and (b)).

   The efficiency is significantly improved compared with a prime focus antenna of the same size, and can reach 65 to 75%. The shape of the aperture of the mirror is generally an oval, (i.e. longer than it is wide), and the LNB is fixed below the mirror by means of a single arm.

   Another interesting feature of the offset antenna is that the angle between the mirror and the vertical is much smaller than for a prime focus antenna, substantially reducing the possibility of snow retention.

   The offset focus antenna is now the most popular and the least expensive type of satellite antenna for dimensions below 1 metre.

3. The '*Cassegrain*' antenna (from the name of an 18th century astronomer who invented the principle and applied it to a telescope) is derived from the prime focus antenna, whose efficiency it improves by improving the LNB coupling and reducing the shadow (at least for large antenna dimensions).

   The LNB source is placed in a hole at the centre (top) of the main parabolic mirror and receives the incoming beam by means of a small secondary mirror, convex and hyperbolic (see Fig. 3.6) placed between the source and the focus (although near to the focus). The efficiency of this antenna can reach 75 to 80%.

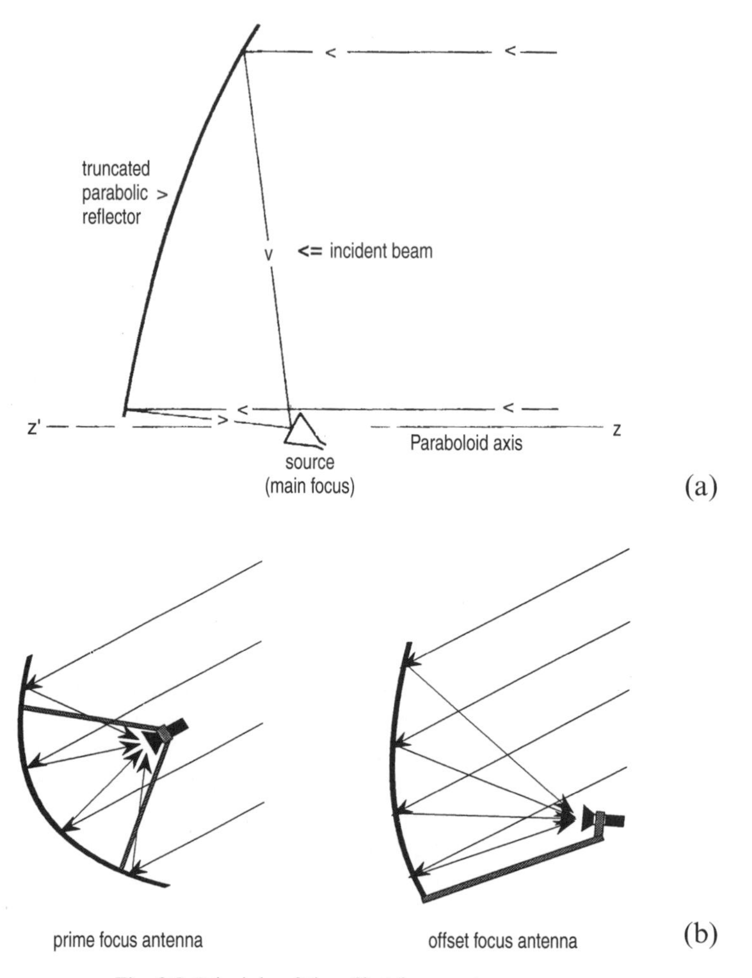

Fig. 3.5 Principle of the offset focus antenna

The Cassegrain antenna is rarely used for consumer antennas (except for some portable versions), but it is the most common form of telecommunications antenna.

4. The '*Gregorian*' antenna is the most sophisticated variant of the parabolic antennas for consumer applications, and it can be considered as a hybrid of the offset focus and Cassegrain antennas. Here, a secondary mirror (concave and parabolic) is placed at the offset focus and the LNB is placed on the same arm of the secondary mirror (Fig 3.7).

   This construction allows an improvement of the LNB coupling as well as completely suppressing the shadow on the main mirror, thus allowing an efficiency that can reach 80%.

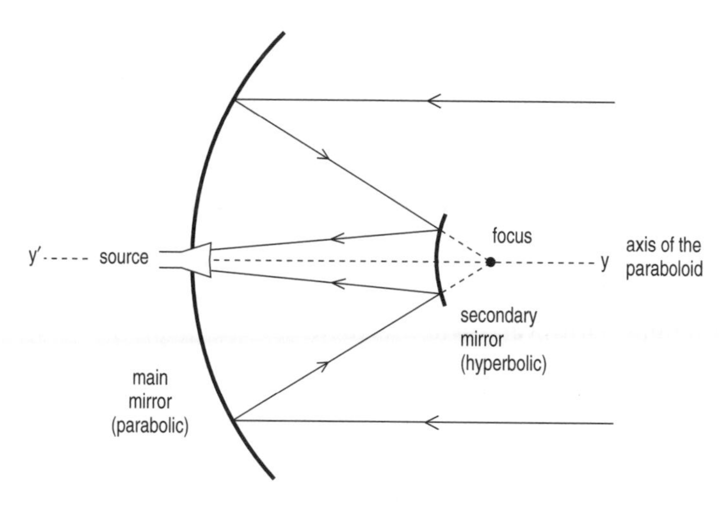

Fig. 3.6 Principle of the Cassegrain antenna

Fig. 3.7 Example of a Gregorian antenna (photo courtesy of Philips)

Owing to its somewhat complex construction (leading to a relatively high price), this type of antenna is not very popular. It is really only justified when the performance increase relative to an offset antenna is of critical importance (for instance when it is required to stay below the limit of 1 m).

3.1.3 Other antenna types

We will briefly mention two other types of antennas, which are not practically used anymore due to their performances not being particularly suited to today's reception conditions and to their cost.

1. The flat antenna is made of a battery of dipoles placed on a dielectric plane and connected in parallel to supply the input of a LNB placed behind this plane. It had its moment of glory at the time of the high power satellites (Marco Polo, TDF-1/2, TV-SAT2) which required only a rather small gain and a limited bandwidth (11.7–12.5 GHz) compared with today's requirements. It allowed realization of square antennas of small dimensions (approximately 40 cm per side) and of very small depth, a distinct aesthetic advantage over the parabolic antenna. However, owing to their limited gain, directivity and bandwidth, such antennas are not well suited to today's conditions (many medium-power satellites with variable spacing in a bandwidth of more than 2 GHz).
2. The horn antenna was proposed during the same period for reception of the same high power satellites. In this case, it is not so much the depth of the antenna that is reduced but its diameter, which can be between 20 and 30 cm, the LNB being placed at the bottom part of the horn. The horn antenna has been no more successful than the flat antenna for the same reasons. We will not consider these two types of antennas again in the rest of the book.

3.2 Source and polarizer

3.2.1 Source

Energy reflected by the parabolic mirror should be transferred as completely as possible to the Low Noise Converter by a device called the '*source*'.

This device is situated at the focus of the mirror and consists either of concentric rings (in the case of prime focus antennas) or of a corrugated conic horn (in the case of offset antennas) followed by an 'adaption waveguide' which transmits the signal to the LNB (see Fig. 3.8).

The role of the source is to adapt the input impedance of the LNB to that of the reflected wave in order to recover the maximum energy at the converter input. The Standing Wave Ratio (SWR) should be as close to 1 as possible (in practice, a value of 1.5 is considered satisfactory).

The connection between the source and LNB is generally by means of a short circular waveguide followed by a circular-to-rectangular transition to adapt the source output to the LNB input (Fig. 3.9).

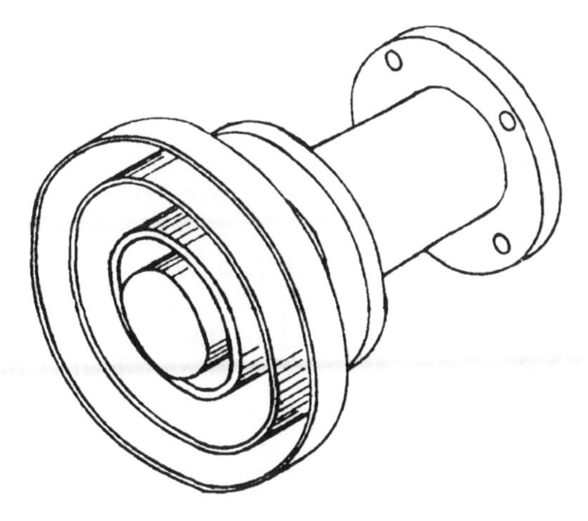

Fig. 3.8 Example of a source with concentric rings

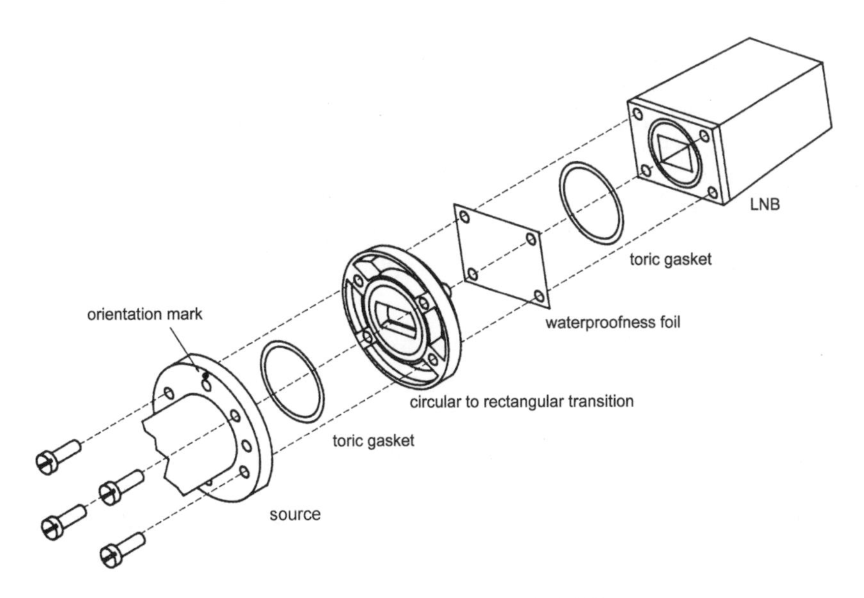

Fig. 3.9 Adaption of the source to the LNB input

The energy of the incoming wave is transmitted to the receiving probe of the LNB, which is made of a quarter wavelength antenna (approximately 0.8 cm for the Ku band). The aperture angle of the source should 'cover' (or 'illuminate') as completely as possible the parabolic mirror's area, but without exceeding it, since this would lead to parasitic directivity lobes that could cause interference from neighbouring satellites or even other terrestrial signals using the same frequency band.

3.2.2 Polarizer

Linear polarization

If the source is connected to the LNB in the way described above, assuming the LNB has only one probe, then only one linear polarization will be received, horizontal or vertical, depending on the orientation of the probe (Fig. 3.10).

With most satellites now using two polarizations (horizontal and vertical), it has been necessary to insert a device between the source and LNB, allowing it to choose the desired polarization through the use of an electrical command coming from the receiver.

The *polarizer* should fulfil this task and, at the same time, ensure a sufficient rejection from the unwanted polarization. The ratio between the desired and undesired polarizations is called the *cross-polarization* attenuation, and should be of the order of 25 dB, at least for a good reception.

Two types of polarizers (magnetic and mechanical) were widely used until the appearance of the integrated LNB (also known as the *Marconi*).

1. The magnetic polarizer is controlled by an adjustable current (up to ± 60 mA) and generates a magnetic field that results in an adjustable rotation of the polarization of the incoming wave.

2. The mechanical polarizer achieves the same result by rotating the receiving probe by means of a small motor controlled by pulses with a variable duty cycle.

These two devices are now mainly used with motorized antennas, in order to cover a substantial part of the geostationary arc, since they allow the polarization to take any intermediate value

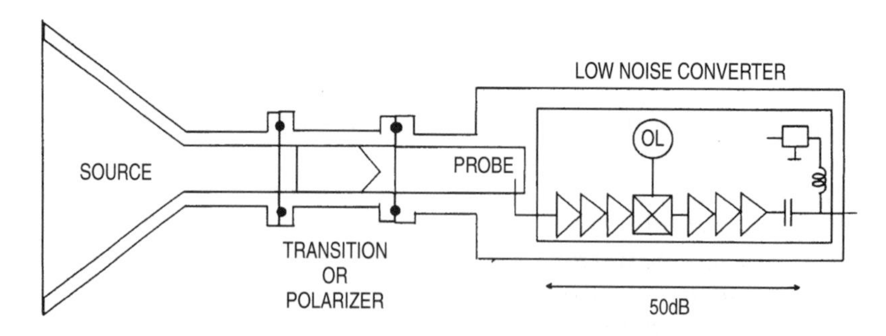

Fig. 3.10 Simplified diagram of the source plus LNB system

between the horizontal and vertical polarizations, which in this case is mandatory. Both require an additional wire to the coaxial cable to control polarization.

The polarization correction, or skew, has to be memorized for each channel being received in order to be automatically controlled at channel change. The integrated LNB, which appeared at the end of the 1980s, drastically simplified the installation of fixed (or motorized antennas, which were intended to cover only a small part of the geostationary arc) by integrating the source, polarizer and low noise converter in a unique waterproof block.

In this case, there is no real polarizer as such, but two orthogonal probes (one horizontal and one vertical) that are switched according to the desired polarization. This switching is controlled by changing the value of the power supply delivered to the LNB by the coaxial cable (13 to 14 V for vertical polarization, 17 to 18 V for horizontal polarization). This avoids the need for a separate cable for the polarizer.

This type of converter in now the *de facto* standard for reception with fixed (offset) antennas. Some versions, without a source, adaptable to a circular guide by means of a standard C120 flange, can be found.

Circular polarization

Circular polarization, which today is in rapid decline in the Ku band, is received in a rather simple way by inserting between the horn and the LNB input a Teflon strip approximately 3 cm long which makes an angle of 45° relative to the receiving probe of the LNB. This strip, or depolarizer, results in a transformation of the incident circular polarization into linear polarization. Depending on the difference in angle (± 45°) of the strip compared with the probe, circular right or circular left polarization is received.

In order to select between the two polarizations, the following is possible:

- either insert a magnetic or mechanical polarizer between the guide containing the strip and the LNB,
- or use an integrated LNB, integrating the Teflon strip in its adaptation waveguide.

Since there are very few LNBs allowing optimal reception of linear and circular polarization, the latter is now often received with a LNB intended for linear polarization, which results in a loss of 3 dB and equivalent reception of both circular polarizations independent of the polarization control (no cross-polarization attenuation). This is generally not too bad due to the high power of the satellites and the relatively small number of remaining channels that can be received (contrary to the case during the period of colocation of TDF1/2 and TV-Sat 2 on 19°W, which used adjacent channels in opposed polarizations).

3.3 The low noise block converter (LNB or LNC)

The Low Noise Block Converter is a key element of the receiving station since the value(s) of its local oscillator(s) and the input filtering define the frequency range being received, while its noise factor (or temperature) will determine the performance of the station. It uses high technology gallium arsenide (GaAs) semiconductors, which, only a few years ago, were reserved by top of the range telecommunication applications (military or civilian). The LNB is made of a receiving probe, a low noise amplifier (LNA), bandpass filters and a frequency downconverter, which lowers the incoming frequency, by means of a local oscillator, into a much lower frequency that is more suitable for transport to the receiver input via a coaxial cable (which can be up to some tens of metres long). This coaxial cable also delivers the power supply (12 to 18 V with a current of the order of 200 mA, depending on the type of LNB).

The LNB electrical gain (between the received power and output power to the cable) is generally between 45 and 55 dB.

In 1998 the noise figure was below 1.5 dB for the very worst Ku band products, and could be as low as 0.6 dB for the 'best-in-class'—the average products being situated between 1.1 and 1.3 dB. The importance of this parameter will be shown later.

3.3.1 Various LNB types

Mechanical variants

Sourceless LNB

The first LNBs did not incorporate the source or the polarizer, to which it was linked by means of a circular to rectangular transition and C120 flange similar to the one used to interconnect rectangular waveguides (see Fig. 3.11).

Fig. 3.11 Example of a sourceless LNB (photo courtesy of Philips)

Fig. 3.12 Example of an integrated LNB (photo courtesy of Philips)

This type of LNB originally covered only one frequency band of less than 1 GHz bandwidth, except for some special double or triple band models. This type of LNB is now practically used only with motorized antennas.

Integrated LNB

At the end of the 1980s, the first models of LNBs incorporating the source and the polarizer appeared; the polarization was controlled by the value of the supply voltage (13 V/18 V). This integrated LNB is often called '*Marconi*', from the name of the company that introduced it.

The so-called 'universal' LNB is an integrated LNB that appeared around 1995 with the first digital TV transmissions on the ASTRA satellites. It allows coverage of the full Ku band in two sub-bands.

Electrical variants

Single band LNB

Until recently, none of the three bands to be received in Europe (Table 3.3) was wider than 800 MHz, and the possibility of receiving more than one band from the same orbital position was very small, so that three variants of single-band LNB have been the *de facto* standard for some time.

Table 3.3 The three bands and corresponding LNBs originally used for Ku-band satellite TV

Band	Frequency (GHz)	Polarization	Satellites (example)	Local oscillator frequency (GHz)	SAT-IF(MHz)
FSS low	10.950–11.700	linear	ASTRA 1A–1C	10.000	950–1700
DBS	11.700–12.500	circular	TDF, TV-SAT, Hispasat	10.750	950–1750
FSS high	12.500–12.750	linear	Telecom 2A/2B/2C	11.475	1025–1275

Only the DBS band was officially allocated to direct-to-home television, and it used circular polarization. The other two bands, called FSS (Fixed Satellite Service), used linear polarization and, in principle, were dedicated to professional applications (programme delivery to cable head-end stations, fixed or occasional transmission between broadcasting sites etc).

Each LNB type was characterized by its local oscillator frequency, chosen in such a way that the LNB output frequency range or SAT-IF (950 to 1750 MHz) was maintained within the input range of the receivers of that time (Table 3.3).

The need to increase the number of available channels was soon recognized on the most popular orbital positions (mainly ASTRA on 19.2°E), and this led to the extension of the low FSS band down to 10.700 GHz (the so-called D band, after the name of ASTRA 1D launched in 1994). The complete coverage of the band has required a definition of a new 'extended FSS' LNB with a lower local oscillator frequency (9.750 GHz) and an extension at the top of the input frequency range of the receivers up to 1950 or 2050 MHz.

A temporary solution, allowing the existing LNB (with a 10.000 GHz oscillator) and receiver to be kept, comprised inserting an up-converter between the LNB and the receiver, which increased, generally by 500 MHz, the SAT-IF corresponding to the D band (from 700–950 MHz to 1200–1450 MHz), bringing these channels within the input range of the receiver through which they were to be received (see Table 3.4).

Table 3.4 The 'extended' FSS (low band) and corresponding LNB

Band	Frequency (GHz)	Polarization	Satellites (example)	Local oscillator frequency (GHz)	SAT-IF (MHz)
Extended FSS	10.700–11.700	linear	ASTRA 1A-1D	9.750	950–1950

'Dual band' (FSS) LNB

For motorized antennas, dual band FSS LNBs have been developed (without either an integrated source or a polarizer). They cover two sub-bands, and switching between the two bands was ensured by means of the supply voltage value (13 V = low FSS band, 18 V = high FSS band), which changed the local oscillator frequency between 10.000 and 11.475 GHz (see Table 3.5)

The DBS band was deliberately ignored owing to the few available transmitters, in circular polarization and D2MAC.

Table 3.5 Characteristics of the dual band LNB (FSS)

Band	Frequency (GHz)	Supply	Local oscillator frequency (GHz)	SAT-IF (MHz)
FSS low	10.950–11.700	13 V	10.000	950–1700
FSS high	12.500–12.750	18 V	11.475	1025–1275

This type of LNB should be used with an external source and a polarizer (electrical or magnetic) suited to the antenna type used (prime focus or offset).

The discontinuation of high power satellites (TDF 1/2, TV-SAT2, etc) operating in circular polarization in the DBS band

(11.700–12.500 GHz) and the introduction of digital television have led, from 1995 onwards, to the use of this band, in linear polarization, at the most popular European orbital positions (19.2°E and 13°E). This has required the extension of the frequency coverage of the LNBs to the whole Ku band, which was made easier by the fact that only linear polarization was required. Two types of LNBs allow this seamless coverage.

Triple band (or full band) LNB

This is simply an extension of the range covered by the dual band LNB, thus allowing full Ku band coverage, (10.700 to 12.750 GHz) which is split into two sub-bands controlled by the supply voltage value (13 V = low band, 18 V = high band). An external source and a polarizer are required for this. Local oscillator frequencies are generally 9.750 and 10.750 GHz (some older types used values of 10.000 and 11.000 GHz but did not cover the part below 10.950 GHz) (see Table 3.6).

Table 3.6 Characteristics of the 'full' or 'triple' band LNB

Band	Frequency (GHz)	Supply	Local oscillator frequency (GHz)	SAT-IF (MHz)
Low	10.700–11.700	13 V	9.750	950–1950
High	11.700–12.750	18 V	10.750	950–2000

'Universal' LNB

For fixed installations, the need to cover the full Ku band has led SES-Astra to define a new type of LNB, known as 'universal', allowing full Ku band coverage (FSS low, DBS and FSS high) in two sub-bands.

It is an integrated LNB (i.e. including source and polarizer) in which polarization is selected by the supply voltage value (13 V = vertical, 18 V = horizontal). Band switching is controlled by means of a 22 kHz signal (0.65 V_{PP} amplitude) which is superimposed on the supply voltage when the high band is selected.

Local oscillator frequencies are 9.750 GHz for the low band and 10.600 GHz for the high band. This has again required the extension of the input frequency range of the receiver to 2150 GHz. The value of 10.600 GHz has been preferred to 10.750 GHz (which is apparently more logical, since it would avoid the extension of the SAT-IF range above 2000 MHz, thus limiting losses in cables) (see Table 3.7).

Table 3.7 Characteristics of the universal LNB

Band	Frequency (GHz)	Satellites (example)	22 kHz	Local oscillator frequency (GHz)	SAT-IF (MHz)
Low (extended FSS)	10.700–11.700	ASTRA 1A–1D	off	9.750	950–1950
High (extended DBS)	11.700–12.750	ASTRA 1E–1G	on	10.600	1100–2150

Figure 3.13 shows the block diagram of the universal LNB.

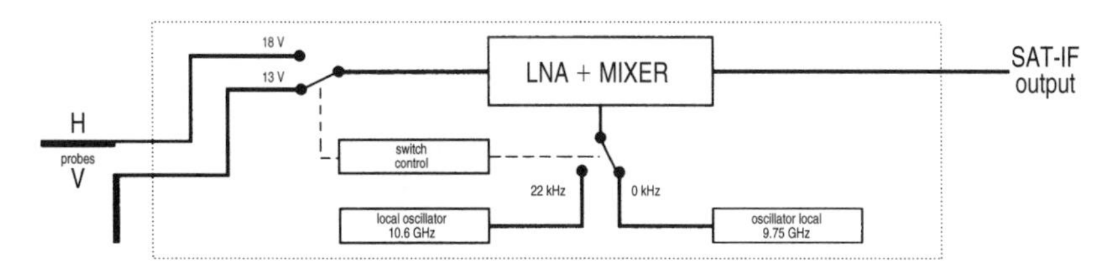

Fig. 3.13 Block-diagram of the universal LNB

3.3.2 Other LNB types

In order to fulfil some specific market requirements, other types of LNBs have been developed

Twin LNB

This type of LNB is in fact made of two electrically independent universal LNBs sharing a common case, source and common probes, with each LNB having its own coaxial output (Fig 3.14). It is of particular interest for residences that, for example, comprise two flats, the inhabitants of which (assumed to be on good terms) do not wish to multiply unduly the number of parabolic dishes on the house.

Another important application is for use in individual houses where two receivers are to be used completely independently (i.e. each being able to receive a TV programme not necessarily in the same band and polarization as the other).

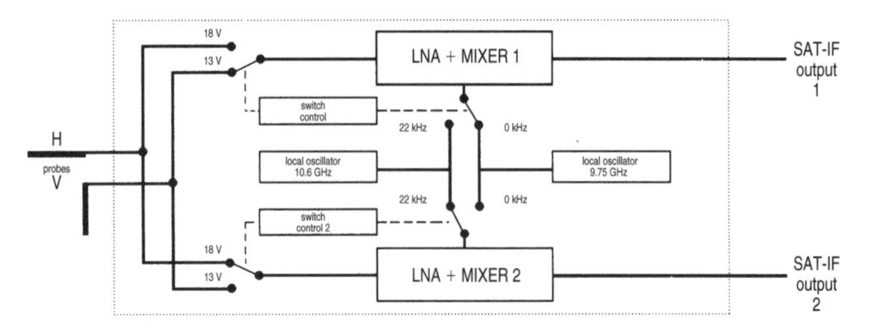

Fig. 3.14 Principle diagram of the twin LNB

'Quad' LNB (or Quattro)

These products are mainly intended for community reception with more than two users. Each contains four independent LNBs in the same housing, and sharing the same source and probes, with each having its own output (Fig. 3.15).

In this type of LNB, there is no switching function since each LNB receives only one band from one polarization, which is permanently available on one of the four outputs:

- low band, vertical polarization,
- low band, horizontal polarization,
- high band, vertical polarization,
- high band, horizontal polarization.

The supply voltage can generally be provided by any one of the four outputs.

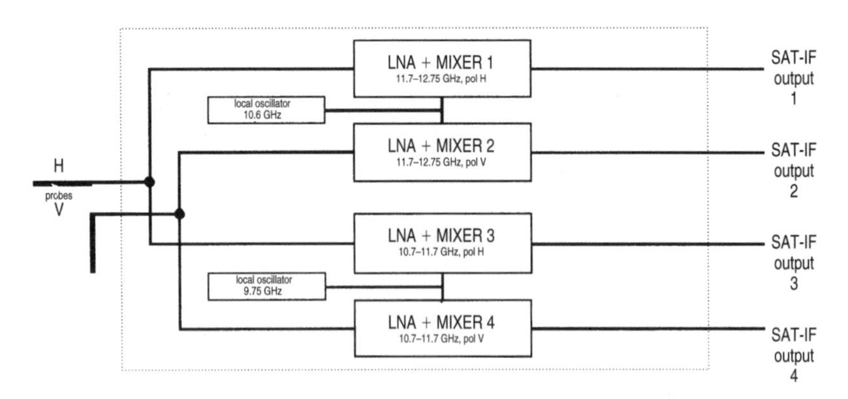

Fig. 3.15 Principle diagram of the 'quad' LNB for switched SAT-IF distribution

Signal distribution requires four coaxial cables connected to a distribution device that ensures the power supply of the LNB, signal amplification and band/polarization selection by each individual user. Each receiver is connected by a unique cable to this device and can select the band by means of the 22 kHz signal, and the desired polarization by means of the supply voltage (13/18 V) in exactly the same way as in the case of individual reception.

Special combined LNB

For the sake of completeness, we will mention some (not very widespread) combined LNBs with a dual horn. Models exist, mainly with an offset of 6°, for reception of ASTRA 1 (19.2°E) and Eutelsat (13°E) with or without an output switch.

This kind of product, with non-adjustable spacing between the two LNBs, only gives satisfactory results for the precise focal distance for which it has been designed, and often requires a special fixing.

A special model (Fig. 3.16) has been developed for reception of Telecom 2A (8°W) and 2B (5°W): since these satellites occupy only 250 MHz of bandwith each (12.500–12.750 GHz), it is possible to have a common parallel output for the two LNBs and to consider these two satellites as if they were only one, by using different local oscillator frequencies (e.g. 11.000 and 11.475 GHz) for the two LNBs.

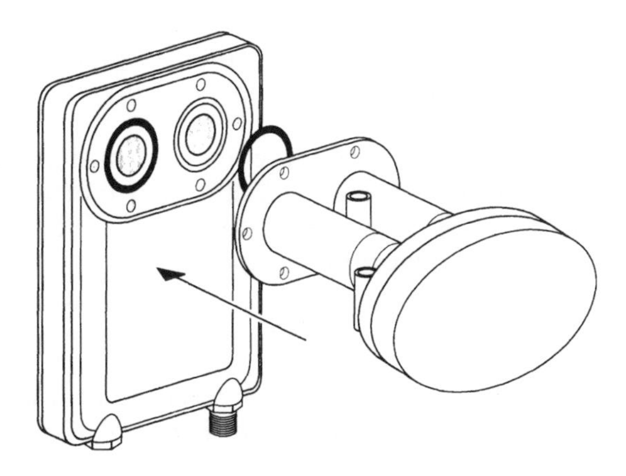

Fig. 3.16 'Hydron' combined LNB for Telecom 2A/2B (courtesy of Philips)

3.4 Link budget

The (down)*link budget* is the series of calculations that allows, by means of the EIRP of the satellite(s) to be received, determination of the required characteristics of the receiving installation (in practice, mainly the antenna dimensions) as a function of the minimum SNR to be guaranteed.

3.4.1 Calculation of the received power by means of the EIRP

The example used in Chapter 1 assumed, for simplification, that there was no energy loss due to transmission (isotropic attenuation A_i = 205 dB at the equator and approximately 206 dB in Europe, at 12 GHz) and a receiving antenna efficiency of 100%.

In practice, we will have to take into account the attenuation—due to the atmosphere—which can vary from 0.5 dB in very clear weather to 8 dB (exceptional precipitation); however, this attenuation does not exceed 1.5 dB for 99% of the worst month at European latitudes. This means that the practical attenuation will be between $A_{i\ min}$ = 206.5 dB (clear weather) and $A_{i\ max}$ = 214 dB (exceptionally heavy rains), but will stay below $A_{i\ ref}$ = 207.5 dB for 99% of the worst month, or 99.75% of the time.

In addition, for a practical installation, we shall have to take into account the efficiency of the antenna. This will lie between 55 and 80% depending on the type of antenna, corresponding to a loss of 2.5 and 0.7 dB respectively; other losses (e.g. LNB coupling and pointing inaccuracy) can be globally estimated to be 1.5 dB.

In the case of a prime focus antenna of 1 m diameter (gross gain 41.9 dB, see Table 3.1) and 55% efficiency (corresponding to a loss of 2.5 dB), the net gain will be, after taking into account the 1.5 dB loss due to the inaccuracy in pointing

$$G_{R\ min} = 41.9 - 2.5 - 1.5 = 37.9 \text{ dB}$$

The minimum power $p_{R\ min}$ (dBW) received by the LNB for 99.75% of the time will then be, for an EIRP of 57.1 dBW.

$$= p_{R\ min} = \text{EIRP} - A_{i\ ref} + G_{R\ min} = 57.1 - 207.5 + 37.9$$

$$= -112.5 \text{ dBW}$$

Let us now take a more representative case of the current receiving conditions: satellite EIRP = 52 dBW, reception with a 80 cm offset dish.

The gross theoretical gain is 40 dB (see Table 3.1) and efficiency can reach 75%, corresponding to a loss of 1.5 dB, to which the coupling and pointing losses (1.5 dB) have to be added.

The net gain is in this case:

$$G_{\mathrm{R\ min}} = 40.0 - 1.5 - 1.5 = 37.0 \text{ dB}$$

and the minimum received power

$$p_{\mathrm{R\ min}} = 52 - 207.5 + 37.0 = -118.5 \text{ dBW}$$

3.4.2 Signal-to-noise ratio or carrier-to-noise ratio (C/N or CNR)

The signal-to-noise ratio is the ratio between the power of the received signal P_{R} and the noise power P_{N} in the considered bandwidth B (in practice, the channel width).

For FM and QPSK modulations used in satellite TV, the amplitude is constant. Thus, the power of the received signal may be considered as equal to the carrier power. The signal-to-noise ratio, SNR, can therefore be assimilated into the carrier-to-noise ratio C/N (S/N and C/N are used interchangeably). The C/N is generally expressed in dB, thus:

$$\mathrm{C/N} = 10 \log (P_{\mathrm{R}}/P_{\mathrm{N}}) = 10 \log (P_{\mathrm{R}}) - 10 \log (P_{\mathrm{N}})$$

It is possible to calculate the C/N in dB directly from the powers expressed in dBW by assuming: $p_{\mathrm{R}} = 10 \log (P_{\mathrm{R}})$ and $p_{\mathrm{N}} = 10 \log (P_{\mathrm{N}})$; then

$$\mathrm{C/N}\ [\mathrm{dB}] = p_{\mathrm{R}} - p_{\mathrm{N}}$$

3.4.3 Noise temperature and noise figure

The *noise temperature* and *noise figure* of a signal source of impedance R (supposed to be a pure resistor) derive from the comparison of its noise power to the noise power of a resistor of the same value (which is proportional to its absolute temperature T in kelvin).

Noise temperature and noise figure of a LNB

The noise factor N of a LNB is the ratio between the noise power produced by the LNB (P_{NL}) and the power (P_{NR}) generated by a pure resistor of the same value and at the same temperature (the standard reference value is 290 K corresponding to 17°C); so $N = P_{NL}/P_{NR}$.

The noise figure NF is normally expressed in dB, so

$$NF\ [\mathrm{dB}] = 10 \log N = 10 \log (P_{NL}/P_{NR})$$

In an ideal case, P_{NL} and P_{NR} would be equal, so $N = 1$ and $NF = 0$ dB.

The 'intrinsic' noise contribution of the LNB (i.e. additional to its output impedance) can then be represented by $N-1$, which allows us to define an 'equivalent noise temperature' at 290 K

$$T\ [\mathrm{K}] = 290\ (N - 1)$$

if we express it from the noise figure, NF, in dB

$$T\ [\mathrm{K}] = 290\ (10^{NF/10} - 1)$$

A noise factor $N = 2$ (noise figure $NF = 3$ dB) is equivalent to a noise temperature of 290 K.

Table 3.8 gives the correspondence between the noise figure and the noise temperature for the most common LNBs for the Ku band (noise figure between 0.7 and 1.5 dB).

Table 3.8 Correspondence between noise temperature and noise figure

NF_{LNB} (dB)	0.7	0.8	0.9	1.0	1.1	1.2	1.3	1.4	1.5
T_{LNB} (K)	50.7	58.6	66.7	75.0	83.5	92.3	101.2	110.3	119.6

It is common usage to specify the noise figure for the Ku band LNBs, but the noise temperature (of the order of 20 to 30 K) is preferred for the C band LNBs, since the corresponding noise figure would be very low.

System noise temperature (T_S)

The noise source of a receiving installation is mainly located in its first stage. In the case of satellite reception, in practice it is the noise coming from the antenna (mainly due to the LNB and to the sky noise and other secondary noise sources received by the antenna).

The noise temperature of the reception system (T_S) is the sum of the noise temperature of the LNB (T_{LNB}) and the antenna noise temperature (T_{ant}). The antenna temperature itself is the sum of the temperature of the part of the sky at which it is pointed (T_{sky}) weighted by the antenna efficiency η (ideally 1) and the contribution from the parasitic directivity lobes (of which a part is directed at the sky, and another part at the Earth) as well as other noise sources: the average of these temperatures T_{ave} has to be weighted by a coefficient Δ (ideally 0) depending on the importance of the parasitic lobes, coupling losses etc.

The system temperature can thus be expressed as

$$T_S = T_{LNB} + T_{ant} \text{ (with } T_{ant} = \eta\ T_{sky} + \Delta\ T_{ave})$$

At 12 GHz, for an elevation of the satellite above 10°, the sky noise temperature (which decreases as elevation increases) is of the order of 30 K in clear weather; it increases greatly when atmospheric conditions are bad and can reach or exceed 270 K in heavy rain. The Earth's noise temperature depends on the environment, but it is often evaluated at 150 K.

In these conditions, with an elevation of 30°, the antenna temperature T_{ant} of a good Ku band offset antenna of 80 cm diameter is of the order of 55 K in clear weather; it can, however, reach 165 K in conditions of very heavy rain.

3.4.4 Figure of merit *G*/*T*

We have seen earlier that C/N is the ratio between the received power P_R, (proportional to the EIRP and the reception antenna gain G_R) and the noise power P_N at the input of the receiving system.

The noise power is proportional to the system noise temperature T_S through the relation

$$P_N\ [W] = kT_s\ B$$

where k is the Boltzmann constant (1.38×10^{-23} W/HzK)

T_s is the system temperature (in Kelvin)

B is the bandwidth of the receiver (in Hz)

Expressed in dBW this gives:

$$p_N \text{ [dBW]} = 10 \log (kT_SB) = 10 \log (kB) + 10 \log T_S$$

we have seen in Chapter 1 that:

$$p_R \text{ [dBW]} = \text{EIRP} - A_i + G_R \qquad (1.5)$$

so the signal-to-noise ratio S/N or C/N can be written

$$\text{C/N [dB]} = p_R - p_N = \text{EIRP} - A_i - 10 \log (kB) + G_R - 10 \log T_S$$

If we let G/T [dB] = $G_R - 10 \log T_S$, the signal-to-noise ratio can be written

$$\text{C/N [dB]} = G/T + \{\text{EIRP} - (A_i + 10 \log kB)\} \qquad (3.1)$$

For a given satellite, at a given receiving position and in given meteorological conditions, the term $\{\text{EIRP} - (A_i + 10 \log kB)\}$ is constant, let it be κ; so we can write: C/N [dB] = $G/T + \kappa$.

The ratio G/T (between the reception antenna gain and the system noise temperature) characterizes the performance of the receiving station; for this reason, it is called the *figure of merit* of the station.

Notes

- The term κ can be positive or negative depending on the EIRP and the atmospheric conditions.
- Although the signal-to-noise ratio (S/N or C/N) is directly dependent on G/T, these two values should not be confused, since they express different notions.

4 Analogue satellite TV

4.1 Analogue video standards (NTSC, PAL, SECAM, D2MAC, PAL+)

We will briefly recall the main characteristics and the differences between the main video standards used in analogue satellite TV. These standards are all based on the compatibility with monochrome television and transmit a video *composite* signal that combines, from one side, synchronization and luminance Y, which is a linear combination of the red, green and blue colour components ($Y = 0.59\ G + 0.30\ R + 0.11\ B$) and, from the other side, the chrominance C carried by a subcarrier located in the high portion of the video spectrum. This method of combining the Y and C signals is called frequency multiplexing. Y represents the brightness of the compatible monochrome picture, independently of the standard. The NTSC, PAL and SECAM systems are different from each other:

- on one hand by the scanning lines number, the frame frequency and the video bandwidth of the monochrome standard on which they have been added (525 lines/60 Hz/4.2 MHz for NTSC, 625 lines/50 Hz/5 to 6 MHz for PAL and SECAM),
- on the other hand, through the method by which the chrominance subcarrier is modulated, as well as its frequency.

The chrominance is made out of two 'colour difference' signals ($R–Y$ and $B–Y$) which represent the colouration of the monochrome picture; the chrominance signals have a reduced bandwidth (of the order of one quarter of the luminance bandwidth).

It is easy to rebuild the RGB signal in the receiver by a linear combination of the *R–Y* and *B–Y* signals and the *Y* signal by a process called 'matrixing'.

These three systems suffer to varying degrees from some artefacts inherent in the principle of frequency multiplexing within the same frequency band, mainly the 'cross-colour' which occurs when luminance frequency components fall within the chrominance band (e.g. the striped shirt of a TV presenter) and a relatively low definition for the chrominance compared with the luminance.

4.1.1 NTSC (USA, 1952)

Historically, the NTSC system (National Television Standard Committee) was the world's first colour television system put into commercial use. In this system, the colour information is carried by two signals, I (In phase) and Q (Quadrature), which are linear combinations of the *R–Y* and *B–Y* signals that modulate a subcarrier at 3.579 745 MHz in quadrature (**QAM**, Quadrature Amplitude Modulation) in order to stay within the video bandwidth of the American TV system (4.2 MHz).

From this modulation process results a vector (Fig 4.1), whose amplitude represents the saturation (or colour intensity) of the picture and whose phase represents the tint (or hue).

In order to allow synchronization of the chrominance demodu-

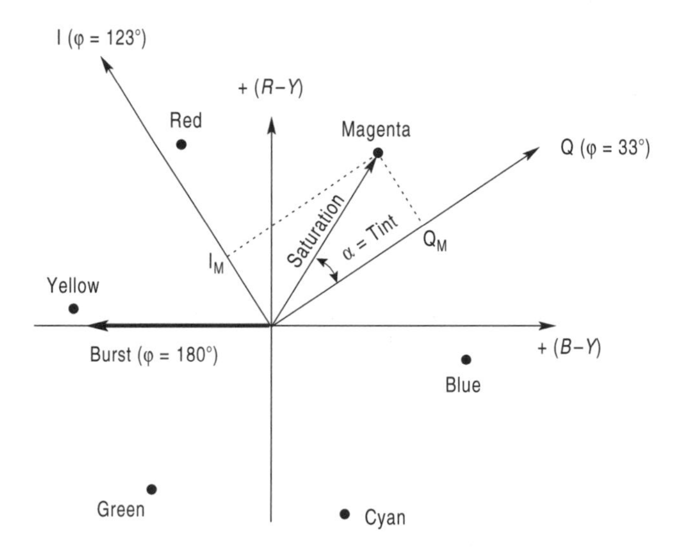

Fig. 4.1 The NTSC colour vector

lator in the receiver, a reference burst of the unmodulated subcarrier is added at the beginning of each line on the black level reference period. The major drawback of NTSC is its extreme sensitivity to phase rotations introduced by terrestrial transmission (the reason for which NTSC has been described as 'Never Twice the Same Colour'). Europeans have tried to find solutions to this problem with the SECAM and PAL systems. The NTSC system is mainly used in the USA, Canada and Japan and in some other countries where the mains frequency is 60 Hz.

4.1.2 SECAM (France, 1957)

The SECAM (SEquentiel Couleur A Mémoire) was the first system that proposed a solution to the hue errors of NTSC by using frequency modulation (FM) for the chrominance—a modulation insensitive to phase rotations.

However, unlike QAM, FM does not allow simultaneous modulation by two independent signals. In order to circumvent this problem, Henri de France (the SECAM inventor) had the simple but ingenious idea of considering two successive lines of a picture as being sufficiently similar as to be considered identical: this allowed alternate transmissions of only one of the two chrominance signals every line; the receiver having the possibility to recover simultaneously the chrominance of the current line and that of the previous line—by means of a delay line, of the duration of exactly one line (64 μs)—and to direct them on the correct output by means of a 'permutator' circuit.

SECAM transmits effectively two colour difference signals D'_R and D'_B alternately (respectively $R-Y$ and $B-Y$ weighted by a different coefficient). The frequency of the unmodulated subcarrier ('rest frequency') is 4.250 MHz for D'_B and 4.406 MHz for D'_R. The frequency change from line to line is used for colour identification and permutator synchronization.

The SECAM system has a high colour rendering accuracy in large areas, even in difficult conditions, but suffers some specific drawbacks due to FM having a limited bandwidth, i.e. poor reproduction of big colour transients and the impossibility of directly mixing two SECAM signals.

SECAM is used in France as well as in its former colonies, in countries of the former Eastern Bloc as well as some Arabic countries.

4.1.3 PAL (Germany, 1961)

The PAL system (Phase Alternating Line) has, in a way, taken 'the best of both worlds' by using the QAM modulation of NTSC and the idea of the SECAM delay line in order to correct its colour errors.

In order to avoid simultaneously the drawbacks of NTSC due to phase rotations and those of SECAM due to FM, the PAL inventor (Walter Bruch) has reused the idea of SECAM, i.e. considering two successive lines as nearly identical; however, maintaining two simultaneous chrominance signals $U = 0.493\ (B-Y)$ and $V = 0.877\ (R-Y)$ which modulate in quadrature (QAM) a subcarrier at 4.433 619 MHz.

In order to allow the correction of phase rotations, the phase of the V signal is reversed every second line; a 64 μs delay line and an adder in the receiver allow addition of the V signal coming from two successive lines, and thus cancel the effect of possible phase rotations. A subcarrier burst on the black level reference, the phase of which alternates between +135 and −135° allows colour identification as well as demodulator synchronization.

The PAL system is used in most European countries as well as in many other countries with a mains frequency of 50 Hz.

4.1.4 D2MAC (Europe, 1985)

In order to improve the quality of pictures and sound from Direct Broadcast Satellites and to try to obtain a unique European standard, the European Community financed the development of a completely new standard, known as D2MAC (Duobinary-Multiplexed Analogue Components). The main technical contributors were the BBC (United Kingdom) and the CCETT (France).

In this system, where the video signal remains analogue but the sound becomes digital, the frequency multiplexing of luminance, chrominance and sound is replaced by a temporal multiplexing, completely avoiding any mixing of signals and, as a consequence, of cross-colour and cross-luminance effects. In addition, the bandwidth ratio between chrominance and luminance is increased from 1/4 to 1/2 and the luminance bandwidth reaches 5 MHz. The system is also designed to support and allow signalling of the picture format (aspect ratio 4/3 or 16/9).

Digital stereo or multilingual sound is supposed to bring 'laser quality', which is not completely true, owing to a lower sampling frequency (32 kHz instead of 44.1 kHz) and to a quantification on 14 bits instead of 16 bits for the Compact Disc. In order to be transmitted sequentially, the video signal has to be time-compressed (in the ratio 2/3 for the luminance signal and 1/3 for the chrominance, which is, in addition, transmitted alternately on every second line as in SECAM) in order to occupy the 52 μs reserved for the 'useful' video in the composite systems.

The digital data carrying sound (eight mono or four stereo tracks), teletext and conditional access data from the Eurocrypt system are transmitted in packets with a duobinary modulation at a bitrate of 10.125 Mb/s during the 12 μs normally dedicated to the synchronization period in composite systems (Fig. 4.2).

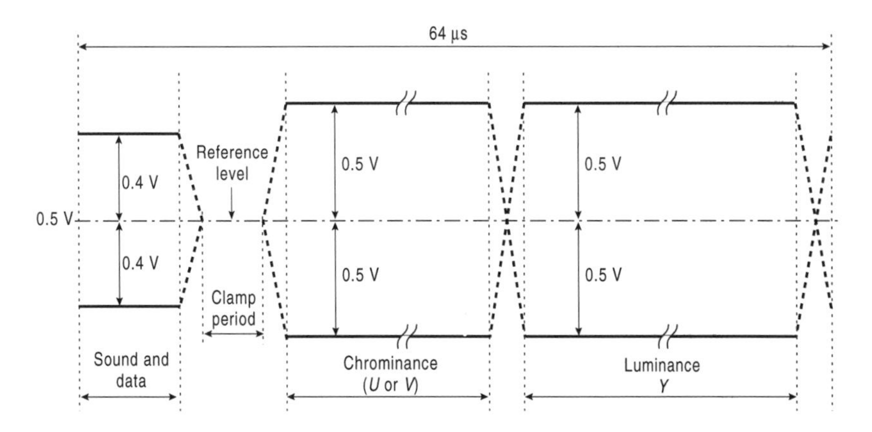

Fig. 4.2 Composition of a D2MAC line

The resulting bandwith is of the order of 8 MHz, but D2MAC can accommodate 7 MHz channels with only a slight loss of video resolution.

The video quality obtained with D2MAC is very satisfactory, but this system has not really developed due to problems with the high power satellites that were carrying it, and also because of the much quicker than anticipated development of digital television.

In practice, D2MAC is used today more as an encryption system rather than a TV standard, and only by Scandinavian satellite TV and French cable TV operators.

4.1.5 PAL+ (Germany, 1990)

PAL+ is an enhancement to PAL, designed in order to allow transmission of pictures in 16/9 aspect ratio, of equivalent quality to D2MAC and to remain 100% compatible with existing PAL receivers. In order to achieve this goal, the 576 active lines of the original 16/9 picture are interpolated to obtain a picture in letterbox format with 432 lines, and the remaining 144 black lines situated above and below the letterbox picture are used to transmit the additional information on the 4.43 MHz PAL subcarrier that is required to rebuild the full resolution picture on a PAL+ receiver. The disturbance of the 4.43 MHz subcarrier on an ordinary PAL receiver is minimal.

The PAL+ system has been used since 1995 on a significant number of satellite programmes from Germany and other European countries. Its major drawback, in addition to being a latecomer, is mainly the high cost of the receiver, which has to realize complex digital signal processing requiring a significant memory size, making its success rather limited.

4.2 The 'baseband' analogue TV signal

In a radio frequency transmission system, the form of the signal to be transmitted before modulation (or after demodulation) is known as the *baseband* signal, independently of its contents, be it analogue or digital. In the case of an analogue signal, this signal is mainly characterized by its waveform and its bandwidth (periodicity, spectrum, DC component etc.).

For satellite analogue television (except in the case of D2MAC, see section 4.1.4), the baseband signal to be transmitted is composed of a composite video signal made by the addition of synchronization, luminance and chrominance signals (see waveform Fig 4.3) as well as one or more associated audio signals.

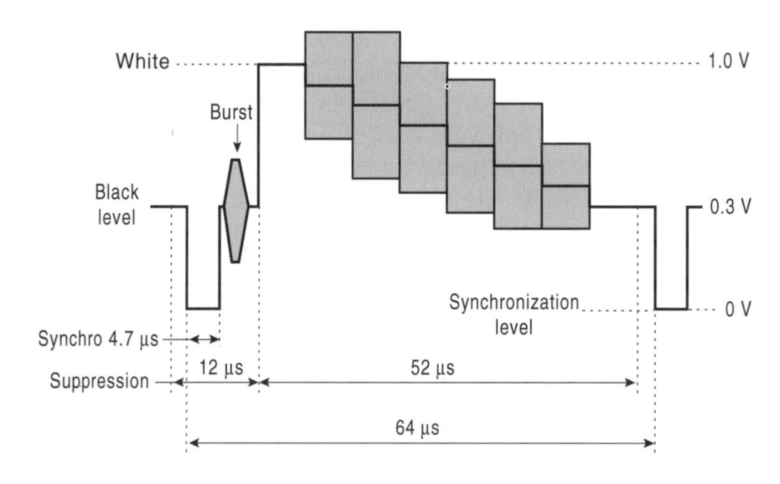

Fig. 4.3 Waveform of a line of a composite video signal

Depending on the TV standard to be transmitted, the video signal bandwidth can theoretically vary between 4.2 MHz (NTSC) and 5 MHz (PAL) or 6 MHz (SECAM). In PAL or SECAM (the case of European satellite TV), the video bandwith is generally limited to 5.5 MHz. The associated audio signal(s) is (are) carried by one or more *subcarriers* placed above the useful video spectrum, i.e. generally between 5.5 and 8.5 MHz. These subcarriers are added to the composite video signal to form the complex baseband signal, the bandwidth of which can exceed 8 MHz (see Fig. 4.4). It is this signal that will modulate the carrier of the satellite transponder.

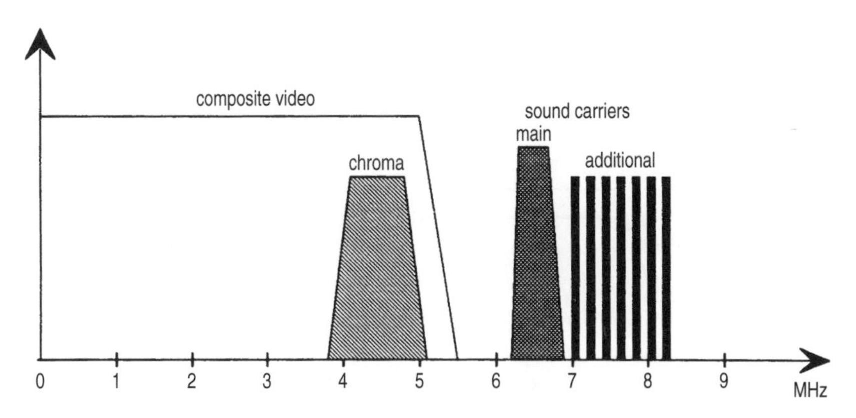

Fig. 4.4 Spectrum of the baseband analogue satellite TV signal

4.2.1 Analogue sound

The audio subcarriers are always frequency modulated (FM) with modulation parameters (subcarrier frequency, FM deviation, audio pre-emphasis) which can vary from one satellite to another, and even between transponders of the same satellite.

The maximum deviation is generally ±75 kHz for the 'main' mono audio subcarrier (200 to 300 kHz bandwidth) and ±50 kHz the 'secondary' audio channels (130 kHz bandwidth).

As in FM radio, the audio signals follow a *pre-emphasis* step that consists of increasing the level of the high frequency components of the audio signal before FM modulation, in order to improve the signal-to-noise ratio for the listener after the reverse operation (de-emphasis) in the receiver. There are four main audio pre-emphasis/de-emphasis 'norms' used in satellite TV, which are differentiated from each other by their response curves and their efficiency in terms of signal-to-noise improvement:

- two of them are simply defined by the time constant of the pre-emphasis network (50 μs or 75 μs),
- the third one, more complex (J17) is named after the number of the CCITT recommendation that specifies it,
- the fourth and last one (known as Wegener Panda or WP), is a so-called 'adaptive' pre-emphasis, which means that it is both a function of the frequency and of the level of the audio input signal (Fig. 4.5); it allows an important improvement of the signal-to-noise ratio.

(Panda is a registered trademark of the American company Wegener Communications, Inc.)

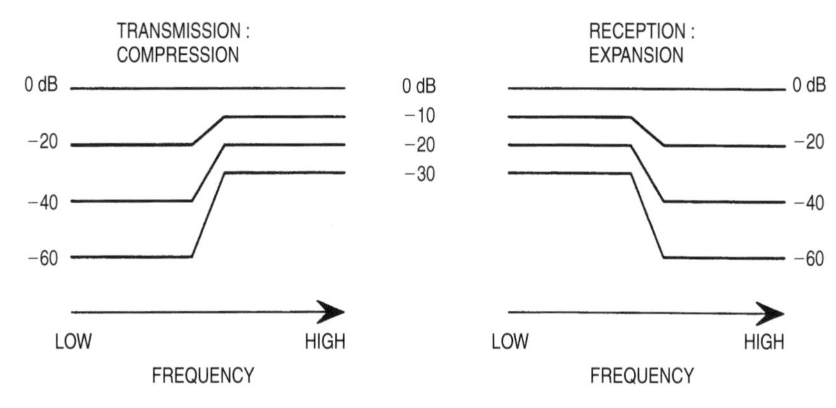

Fig. 4.5 Principle of adaptive pre-emphasis

Table 4.1 summarizes the main characteristics of analogue audio subcarriers used in Europe.

The most common audio configuration for analogue TV sound is a 'main' subcarrier at 6.50 or 6.60 MHz with 50 μs pre-emphasis, accompanied by a certain number of 'auxiliary' sub-carriers between 7.02 and 8.28 MHz (see Fig 4.4 and Table 4.1) separated by 180 kHz from each other and using either a fixed 75 μs pre-emphasis or an adaptive Wegener Panda pre-emphasis.

Most of the time, the mono audio sound is duplicated in stereo on the 7.02 + 7.20 subcarrier couple, and the subcarriers from 7.38 MHz and above are used to transmit audio channels not linked to the TV programme (most often stereo radio channels). Alternatively, some TV programmes transmit many associated audio channels in different languages and, for this purpose, use all or part of the subcarriers between 7.02 and 8.28 MHz.

Table 4.1 The most usual parameters used in Europe for analogue audio subcarriers

Satellites (examples)	Frequency (MHz)	Deviation (kHz)	Pre-emphasis	Remark
Telecom 2B	5.80	±75	J17	purely French
Astra 1A–1D	6.50	±75	50 μs	absent on ADR TXP
Eutelsat, TC2A	6.60	±75	50 μs	the most common
Eutelsat	6.65	±75	50 μs	less common
Astra, Eut., TC2A	7.02 + 7.20	±50	75 μs or WP	stereo TV sound
Astra, Eutelsat	7.38 + 7.56	±50	75 μs or WP	stereo radio
Astra, Eutelsat	7.74 + 7.92	±50	75 μs or WP	stereo radio
Astra, Eutelsat	8.10 + 8.28	±50	75 μs or WP	stereo radio

4.2.2 Digital sound for TV (NICAM)

Except for D2MAC, digital sound is very seldom used for analogue TV channels, but there have been some cases where **NICAM** has been used with the same parameters as terrestrial transmission (B/G or L systems, see Table 4.2).

Only Telecom 2B has used this standard to supply cable TV head-end stations; this standard did not develop for direct-to-home TV due to the fact that consumer demodulators do not include a NICAM decoder, and since the internal decoder of the TV set cannot be used through the SCART connector, it requires an external NICAM decoder.

Table 4.2 Main characteristics of the NICAM sound used on satellite

TV norm	Subcarrier	Bit-rate	Modulation	Satellite	Remark
NICAM B/G	5.85 MHz	728 kb/s	QPSK	Telecom 2B	now obsolete

4.2.3 Digital radio (ADR)

In 1995, ASTRA launched a digital radio broadcast system (**ADR**: Astra Digital Radio) on analogue satellite TV transponders. This system is based on the MPEG-1 audio (layer 2) norm, also known as MUSICAM, which is also used for the European digital terrestrial radio system DAB (Digital Audio Broadcast) and the digital television system DVB (Digital Video Broadcast). The main characteristics are summarized in Table 4.3.

Table 4.3 Main characteristics of the ADR system

Audio coding	Sampling rate	Bandwidth	Dynamics	Audio bit-rate	Data bit-rate
MUSICAM	44.1 or 48 kHz	20 Hz–20 kHz	> 90 dB	192 kb/s	9.6 kb/s

ADR is mainly used by a subscription music radio network and by German radio broadcasters to replace the defunct **DSR** system (Digital Satellite Rundfunk) which transmitted 16 hi-fi radio programmes per channel on TV-SAT and Kopernikus.

The ADR system uses the same subcarrier frequency grid (with 180 kHz spacing) as the auxiliary FM subcarriers (WP). Their frequency range has been extended by suppressing the 'main' mono audio subcarrier at 6.50 MHz (M*) and by using a lower modulation index (0.12) for the ADR subcarriers. This has allowed the maximum number of subcarriers to increase to 14, ranging from 6.12 to 8.46 MHz, out of which 12 can be modulated in **QPSK** by the ADR signal (with a bit-rate of 192 kb/s for the stereo audio signal and 9.6 kb/s for data used, for instance, to give additional information on the current programme).

The two subcarriers at 7.02 and 7.20 MHz are always reserved for analogue TV sound, and there is always an FSK modulated subcarrier at 8.595 MHz (N_c) dedicated to the network's control.

Table 4.4 shows the five audio subcarrier configurations allowed on ASTRA analogue TV channels.

There are now satellite receivers allowing both analogue satellite TV reception and ADR reception. However, one can have some doubts over the long term future of this system, which is limited to the analogue channels of ASTRA, at a time when radio transmissions using the DVB world standard—which can be received by any DVB compliant digital TV satellite receiver—have started everywhere.

4.3 RF modulation, video pre-emphasis, energy dispersal

As we saw in Chapter 3, due to the relatively low power of the satellite transmitters and the very long journey of the signal, the C/N ratio of satellite reception is rather low, between 10 and 15 dB. Consequently it was then necessary to chose an RF

Table 4.4 Different possible audio subcarrier configurations (from SES/ASTRA)

#	6.12	6.30	6.48	6.66	6.84	7.02	7.20	7.38	7.56	7.74	7.92	8.10	8.28	8.46	8.595
1	–	–	M*	–	–	**WP**	**WP**	WP	WP	WP	WP	–	–	–	$\mathbf{N_c}$
2	–	–	M*	–	–	**WP**	**WP**	WP	WP	ADR	ADR	ADR	ADR	ADR	$\mathbf{N_c}$
3	ADR	ADR	ADR	ADR	ADR	**WP**	**WP**	WP	WP	WP	WP	–	–	–	$\mathbf{N_c}$
4	ADR	ADR	ADR	ADR	ADR	**WP**	**WP**	WP	WP	ADR	ADR	ADR	ADR	–	$\mathbf{N_c}$
5	ADR	ADR	ADR	ADR	ADR	**WP**	**WP**	ADR	ADR	ADR	ADR	ADR	ADR	ADR	$\mathbf{N_c}$

modulation optimizing the signal-to-noise ratio of the demodulated signal.

4.3.1 RF modulation

Numerous subjective tests on a vast number of viewers have shown that a minimum signal-to-noise ratio (S/N), of the order of 45 dB (weighted), is required to obtain a picture quality judged by the viewers as 'excellent'.

The most appropriate modulation, universally used for satellite video broadcasting is frequency modulation (FM). As we will see below, with suitable parameters, it permits a weighted signal-to-noise ratio of the order of 45 dB to be achieved with a satellite signal showing a C/N of the order of 15 dB. Regarding the bandwidth of the modulated signal, the spectrum of an FM modulated signal shows theoretically an infinite number of stripes of decreasing amplitude, each a distance from the previous one of the frequency of the modulating signal. However, it is obvious that the limits of the allocated transmission channel have to be respected; it is fortunately possible to limit to three or four the number of stripes (or *modulation index*) without any consequence other than a slight degradation of the picture contours in order to stay within the limits of the transmission channel.

The modulation 'slopes' used in analogue satellite TV depend on the width of the RF channel (generally between 26 and 36 MHz); the most common values are 13.5 MHz/V (D2MAC), 16 MHz/V (ASTRA) or 22 MHz/V (Eutelsat, TELECOM). This is why it is very useful that the receiver includes a demodulation slope control (at least for multisatellite reception) to obtain a video output voltage near the standard value of 1 V_{pp}, in order to avoid any distortion of the output signal and to have a quasi-constant contrast on every channel of all received satellites (most receivers can switch between two fixed values).

4.3.2 Video pre-emphasis

As for audio signals, in order to improve the signal-to-noise ratio at high frequencies of the video signal—the level of which decreases rapidly with frequency—a video pre-emphasis compliant with the CCIR 405-1 recommendation is applied before RF modulation (see the curve in Fig. 4.6 for PAL or SECAM signals; the curve used in D2MAC is different). This has the effect of increasing the level of the high frequencies relative to the noise floor; their

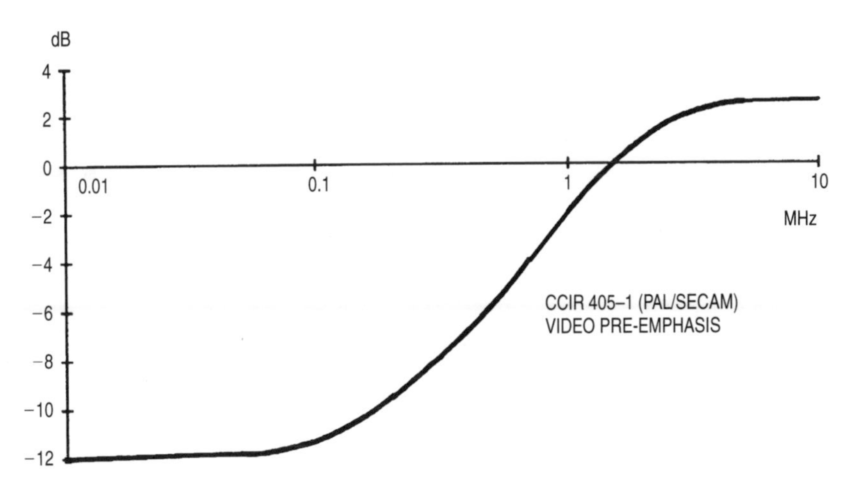

Fig. 4.6 CCIR 405-1 (PAL/SECAM) video pre-emphasis curve

original level is restored in the receiver by a de-emphasis network, which has the reverse frequency response curve, and which consequently reduces the relative noise level at high frequencies, thus improving substantially the quality of the received picture.

4.3.3 Energy dispersal

The video signal is not sinusoidal, and it can have a relatively important DC component that depends on the picture contents (in fact, a grey picture of uniform luminance results in an almost DC signal, only interrupted by the synchronization pulses).

Without any precautions, it would result, after FM modulation, in an RF signal of almost constant frequency containing all the transmitted energy, which could result in interferences with the neighbouring services, especially to uplinks of narrow band telecommunication services). In order to avoid this problem, an '*energy dispersal*' signal, consisting of a sawtooth of 25 Hz, of which the slope is inverted at every field synchronization, is superimposed onto the video signal (Fig. 4.7).

The relative importance of this signal to the maximum video amplitude is not the same for all satellites, and it depends on the channel width; the amplitude of the energy dispersal signal is generally doubled in the absence of a modulating video signal (see Table 4.5). In this way, even in the case of a constant luminance picture, the carrier frequency is always modulated by a 25 Hz signal.

This signal has obviously to be eliminated in the receiver in order to recover the original video signal; this is done in a relatively

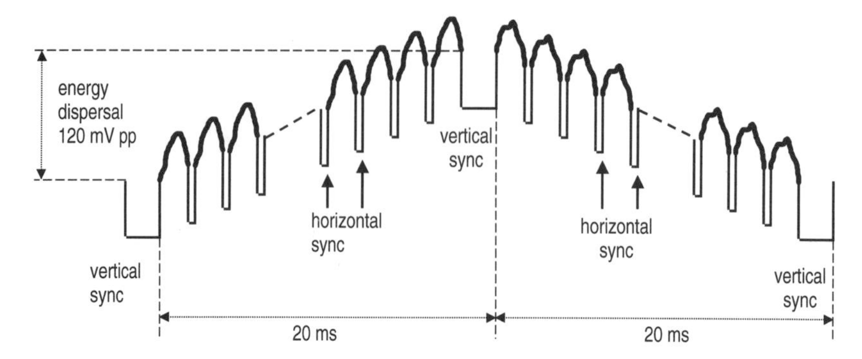

Fig. 4.7 Video signal with (exaggerated) energy dispersal signal

Table 4.5 Modulation parameters of the main European satellites (analogue mode)

Satellite		ASTRA	EUTELSAT		TELECOM
channel width		26 MHz	27 MHz	33–36 MHz	36 MHz
FM deviation		16 MHz/V	16 MHz/V	22 MHz/V	22 MHz/V
Energy	with signal	2 MHz pp	1.2 MHz pp	2 MHz pp	2 MHz pp
dispersal	without signal	4 MHz pp	4 MHz pp	4 MHz pp	4 MHz pp

straightforward manner by means of simple clamp circuitry that aligns the video signal on an internal reference voltage during the black level reference period at the beginning of each line.

4.3.4 Carrier-to-noise ratio (C/N), video signal-to-noise ratio and picture quality

In the case of a carrier frequency modulated by a standard video signal with a ratio of 30% between the synchronization signal and maximum video amplitude (i.e. from sync to white), the relation between the unweighted video signal-to-noise ratio and the C/N ratio is given by:

$$\mathrm{S/N_{uw}} = 3/2\ (\mathrm{C/N}) \times (\Delta F/f_{max})^2 \times (BW/f_{max})$$

where:

ΔF is the maximum FM deviation (e.g. 16 MHz for a 26 MHz channel)

f_{max} is the video bandwidth (say 6 MHz)

BW is the channel width (26 MHz in our example)

which gives: $S/N_{uw} \approx 46 \times C/N$ with the above values, or, in dB:

$$S/N_{uw} = 10 \log 46 + C/N \approx C/N + 16 \text{ dB}$$

A weighting curve has been determined experimentally in order to take into account the subjective disturbance generated by the noise as a function of its spectrum. It brings an improvement of approximately 11 dB for a PAL or SECAM signal, to which the pre-emphasis adds another 2 dB; the resulting weighted S/N ratio is then:

$$S/N_w \approx C/N + 29 \text{ dB (see Note 1)}$$

The relation between the weighted S/N ratio and the subjective impression of picture quality judged by a panel of testers has been translated by the CCIR into a normalized scale with five levels.

5 'excellent' picture; $S/N_p > 45$ dB ($C/N > 16$ dB)

4 'good' picture; $S/N_p > 41$ dB ($C/N > 12$ dB)

3 'correct' picture; $S/N_p \approx 38$ dB ($C/N \approx 9$ dB)

2 'poor' picture; $S/N_p \approx 36$ dB ($C/N \approx 7$ dB)

1 'bad' picture $S/N_p < 35$ dB

4.4 Scrambling and conditional access

More and more TV transmissions are now pay programmes, and their video signal is therefore scrambled; their audio signals are rarely scrambled, however, due to the unavoidable degradation that would result. We will not discuss the (now obsolete) first scrambling systems used in the 1980s (based, for instance, on video polarity inversion or variable line delay).

Three main conditional access systems are now used in Europe for analogue transmissions: Nagravision/Syster and Videocrypt for PAL or SECAM, Eurocrypt for D2MAC transmissions. All three systems use a detachable chip which contains the access rights of the user, which are downloadable by off-air data transmissions from the satellite.

In the case of the Nagravision/Syster system, the chip is contained in a kind of plastic key, whereas the Eurocrypt and Videocrypt systems use a standard ISO 7816 chip card format. In all three systems, the decoders are all equivalent and only the card or key is 'personalized'.

We will give below, for reference only, some 'public domain' features which differentiate these systems.

4.4.1 Nagravision/Syster

This system was developed at the end of the 1980s by the Swiss company Kudelski SA and its first application was the replacement of the first generation Canal+ decoders (Discret system). It was then generalized to all satellite analogue programmes of the Canal+ group, but it is also used on some cable networks and satellites independent of Canal+.

The principle of the video scrambling consists of modifying, in a pseudo-random sequence, the transmission order of 32 lines among the 288 of a TV field (a process known as 'line shuffling'), which makes the picture unwatchable, even if one can vaguely see what is happening on the screen. In addition, since the order of the lines is modified, the colour identification can be affected and the scrambled picture can appear in black and white.

The chip in the key of an authorized user gives to the descrambler the initialization word necessary for the pseudo-random generator to issue the correct sequence. The chip contained in the key, if it has adequate rights, calculates this word (which changes periodically) from the data transmitted at regular intervals by the transmitter during the vertical blanking interval (**VBI**). The rights contained in the key can be activated, modified or cancelled from the data transmitted in the same manner.

The audio channel can, if required, be 'scrambled' in a very rudimentary manner since it is merely a spectrum inversion around a frequency of 12.8 kHz (unchanged from the Discret system). The Nagravision system has resisted piracy for a long time, partly because decoders were not sold in the retail market but rented to subscribers.

4.4.2 Videocrypt

This system was developed at roughly the same time as the previous one by Thomson Multimedia and is mainly used by the analogue BSkyB collection on Astra. The principles of conditional access are very similar to those of Nagravision/Syster, but the key is replaced here by a smart card. The video scrambling, however, is very different.

The principle used here consists of inverting the transmission order of the beginning and the end of each line relative to a point, the position of which varies from line to line (256 possible values) in a pseudo-random manner, which renders the picture completely distinguishable (this process is known as 'cut and rotate').

The chip of the smart card of an authorized user gives to the descrambler the initialization word necessary for the pseudo-random generator to find the correct position of the cutting point. The chip, if it has adequate rights, calculates this word (which changes periodically) from data transmitted at regular intervals by the transmitter during the vertical blanking interval (VBI). The rights contained in the key can be activated, modified or cancelled from data transmitted in the same manner.

Sound is not scrambled in the Videocrypt system, which exists in two variants that differ only by their scrambling algorithms. Since the decoders are available through retail channels, Videocrypt suffered from piracy shortly after its commercial launch. This has led the broadcasters to numerous Electronic Counter Measures (**ECM**) to invalidate pirate cards and even to change the cards of their official subscribers. The latest countermeasures, however, seem to have put an end to large-scale piracy.

4.4.3 Eurocrypt

This system has been developed to be used jointly with the D2MAC standard. The principles of conditional access are similar to those of the two previous systems.

The cut and rotate principle of video scrambling is the same as the one used in Videocrypt, but here one or two cutting points are possible, in order to increase the difficulty of piracy.

Sound, which is digital in the D2MAC standard, can be optionally scrambled without degradation by a simple convolution of the bitstream by a pseudo-random sequence.

The chip of the smart card of an authorized user gives to the descrambler the initialization word necessary for the pseudo-random generators to find the correct positions of the cutting points and to output the right sequence for descrambling the audio bitstream. The chip, if it has adequate rights, calculates this word (which changes periodically) from data transmitted at regular intervals by the transmitter in the duobinary data stream which carries digital sound and conditional access data. The rights contained in the key can be activated, modified or cancelled from data transmitted in the same manner.

Eurocrypt exists in two variants (M and S), which differ only by their encryption algorithms (and by the smart card), but both suffered from piracy a few years after their launch.

The D2MAC/Eurocrypt standard is now only used on satellites in Scandinavian countries and in France on some cable networks with the Visiopass decoder.

4.5 The analogue satellite receiver

From the preceding explanations, the minimal functions that have to be included in an analogue satellite receiver (or tuner) can be deduced, and are illustrated by the block-diagram in Fig. 4.8.

TUNER: This block is located in a shielded box and ensures channel selection, amplification and frequency down-conversion of a channel in the Satellite Intermediate Frequency (between 950 and 2150 MHz) to a standardized IF frequency (480 MHz in Europe). This signal is then filtered with a bandwidth of 27 or 36 MHz by means of a Surface Acoustic Wave filter (**SAW**), amplified and limited and finally FM demodulated. The tuner delivers at its output the non-de-emphasised baseband signal (audio plus video). For an accurate generation of the local oscillator frequency, it includes a frequency synthesizer controlled by the microcontroller of the receiver via a serial bus (I2C or 3 wires bus). The tuner can include one or two SAT-IF inputs.

VIDEO PROCESSING: In the case of a PAL/SECAM or NTSC receiver, the video processing is limited to the rejection of the audio subcarriers, de-emphasis of the video signal (CCIR 405-1), clamping in order to remove energy dispersal, and finally amplification of the signal in order to deliver a standard composite video signal (1 V/75 Ω) to the video input of the TV (SCART or Cinch). PAL, SECAM (or NTSC) decoding is performed by the decoder of the TV set, which must be compatible with the video standards being received.

AUDIO PROCESSING: This part is, in practice, the most complex part of the receiver, taking into account the different possible audio configurations. The various subcarriers between 5.5 and 8.5 MHz are in the high portion of the spectrum of the baseband signal at the tuner output. After high pass and band pass filtering, a frequency conversion generally takes place by means of a frequency synthesizer, programmable between 16.2 and 19.2 MHz

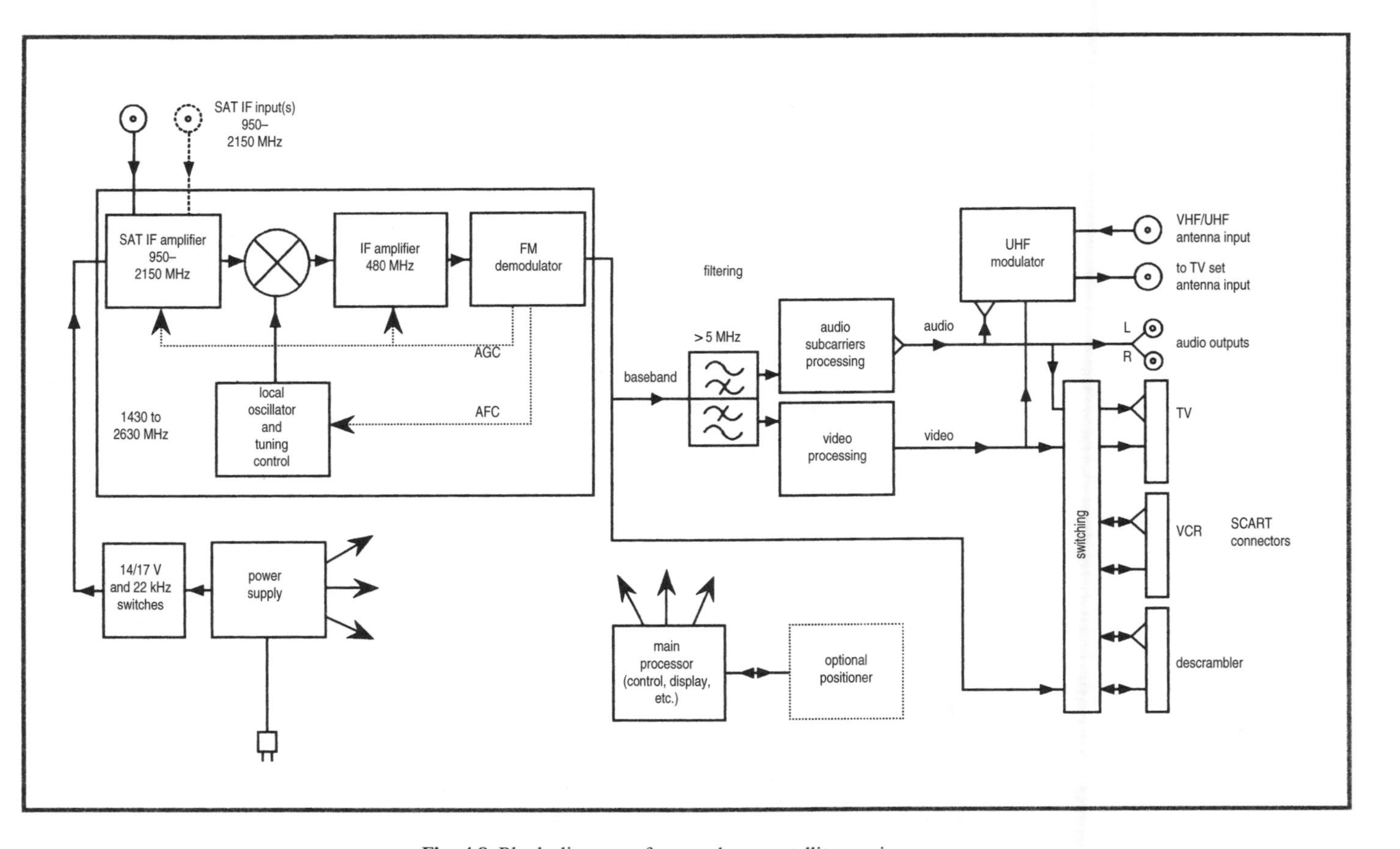

Fig. 4.8 Block diagram of an analogue satellite receiver

via a serial bus by the microcontroller. If the 'main' audio channel has been chosen, a wideband (≈250 kHz) IF signal at 10.7 MHz is available. This signal is filtered, amplified and limited before being FM demodulated and de-emphasized with the appropriate curve (50 μs or J17), and then fed to the audio outputs. If an 'auxiliary' stereo signal is selected, then two identical narrow band (≈130 kHz) IF chains are used, one working at 10.7 MHz and the other at 10.52 MHz (180 kHz spacing). These signals are filtered, amplified and limited before being FM demodulated and de-emphasized with the appropriate curve (75 μs or WP in this case) before being applied to the right and left outputs. A battery of switches allows the choice between the various possible audio options (mono, stereo, language in the case of multiple soundtracks etc.).

CONTROL: The receiver is under the control of a microcontroller which commands all the programmable circuitry (tuner, audio, LNB supply, 22 KHz, optional polarizer and positioner) and stores, in a non-volatile memory (EEPROM), all the parameters of the programmes selected by the user. On recent receivers, which can store many hundreds of programmes, search and store operations are eased by an 'On Screen Display' (**OSD**) or 'On Screen Graphics' system.

CONNECTIONS: Most recent receivers have three SCART connectors (one for the TV, one for the VCR, and one for an external descrambler (Nagravision, Videocrypt or other) and separate Cinch audio outputs. Some also have a UHF modulator to interface older equipment without A/V input and a non-de-emphasized baseband output for an external D2MAC decoder, as well as specific connections for a magnetic or mechanical polarizer, and sometimes even an antenna positioner.

Most receivers sold in the United Kingdom have a built-in Videocrypt decoder (in this case, it is called an Integrated Receiver-Decoder, or **IRD**).

Receivers with a built-in D2MAC decoder, in addition to the above described functions, have been produced, but are now rather seldom used except in the Scandinavian countries. The only common parts of the PAL/SECAM and the D2MAC functionality are the tuner and the power supply.

We will not describe the D2MAC decoder, which is of little interest today, but will simply say that it has for input the non-de-emphasised baseband signal out of the tuner and supplies a RGB

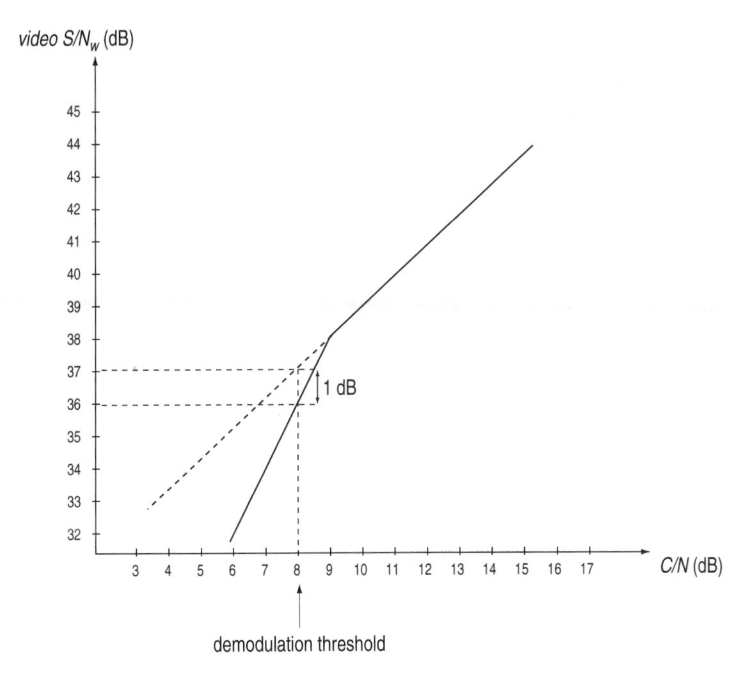

Fig. 4.9 Curve $S/N_w = f\,(C/N)$ illustrating the demodulation threshold phenomenon

plus synchro signal and a stereo audio signal to the SCART connector. In addition it has a PAL or SECAM encoder required for VCR recording, and one (sometimes two) smart card readers for the Eurocrypt conditional access system.

Note 1

When the C/N ratio drops below a certain value, called the '*demodulation threshold*', disturbances known as white or black '*sparklies*' or 'fishes' appear on a picture (especially in the most coloured part of the picture, due to its important high frequency contents) which otherwise would still be of a relatively acceptable quality.

This threshold corresponds to the point where the curve $S/N_w = f\,(C/N)$ is 1 dB below the prolongation of the straight line described by $S/N_w = C/N + 29$ dB (see Fig 4.9).

Below this point, the video signal-to-noise ratio, S/N_w, and the picture quality decrease much more rapidly as a function of C/N.

For common consumer receivers, this threshold corresponds to a value of the order of 8 dB. The value announced for the threshold should be taken cautiously however, since there is no real standard measuring procedure, and so it depends on the picture contents. There are also receivers with so-called 'threshold extension' which sometimes have threshold values as low as 3 or 4 dB; this low value is obtained mainly by reducing the IF bandwidth, at the expense of other artefacts (sparklies on the important luminance transients and in the high saturation parts of the picture, where the high frequency contents are important).

5 Digital satellite television

5.1 Digitization of audio and video signals

For many years, video professionals have been using various digital formats. In order to ease the interoperability of equipment and international programme exchange, the former CCIR (Comité Consultatif International des Radiocommunications—now ITU-R) has standardized conditions of digitization (recommendations CCIR-601 and CCIR-656) of digital video signals in component form (Y C_r C_b in 4:2:2 format).

The main advantages of these digital formats are that they allow multiple copies to be made and editing to be performed without any quality degradation, as well as making international exchanges easier. However, the drawback is a very large bit-rate which makes these formats unsuitable for transmission to the end user without prior signal compression.

5.1.1 The 4:2:2 and 4:2:0 formats

The 4:2:2 format

Recommendation CCIR-601 defines digitization parameters for video signals in component form based on a Y C_b C_r signal in 4:2:2 format (four Y samples for two C_b samples and two C_r samples) with a *quantization* on 8 bits.

The sampling frequency is 13.5 MHz for luminance and 6.75 MHz for chrominance, regardless of the standard of the input signal (525 lines/60 Hz or 625 lines/50 Hz). This results in

720 active video samples per line for luminance, and 360 active samples per line for each chrominance. Chrominance resolution is half the luminance resolution in the horizontal direction, but these are identical in the vertical direction.

The resulting gross bit-rate is then: $13.5 \times 8 + 2 \times 6.75 \times 8 =$ 216 Mb/s. If we consider that it is useless to digitize the line and field *suppression* periods of the signal, the digitization of the useful video part requires 'only' 166 Mb/s.

Recommendation CCIR-656 defines standardized interfacing conditions for exchange of 4:2:2 signals. These signals are digitized according to recommendation CCIR-601 and are provided in a time multiplexed form (C_{r1} Y_1 C_{b1} Y_2 C_{r3} Y_3 C_{b3} . . .) on an 8 bit parallel interface, together with a 27 MHz clock (one clock period per Y, C_b or C_r sample). Synchronization and other data are included in the data flow.

The 4:2:0 format

For less demanding applications in terms of resolution, aiming at transmission bit-rates as low as possible, some by-products of the 4:2:2 format have been defined, out of which the 4:2:0 format has been selected for consumer digital television; it is obtained from the 4:2:2 standard by using the same chrominance to colourize two successive lines, in order to reduce the bit-rate (to 133 Mb/s approximately) and the required memory size in the processing circuits. In addition, it results in an equivalent resolution in the vertical and horizontal directions (equal to half the luminance resolution).

Figure 5.1 shows the position of samples in the 4:2:0 format.

This 4:2:0 format is the base used for D2MAC and MPEG-2 (MP@ML) coding.

The **SIF** format (Source Intermediate Format), which has half the resolution in both directions compared with 4:2:0, is used as a base for MPEG-1 compression.

From the above explanation, it appears that a bit-rate of more than the 100 Mb/s required by the 4:2:0 format 'as is' means that it is not directly usable for broadcasting the signal to the final user, since it would require a bandwidth of many times the one occupied by a PAL or SECAM signal. This is why, in order to be able to start digital television broadcast services, solutions had to be found to problems which can be classified into two categories.

1. *Source coding*: compression techniques used for video and audio signals aimed at reducing as much as possible the bit-

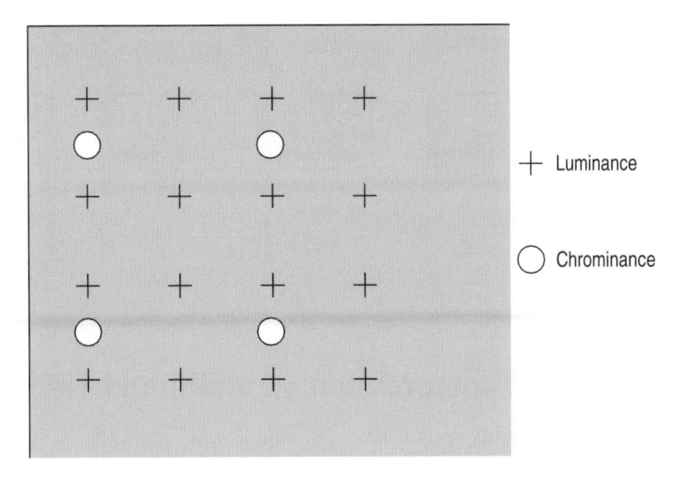

Fig. 5.1 Position of samples in the 4:2:0 format

rate with the lowest perceptible quality degradation of the picture and the associated sound by reducing the *redundancy* of the audio/video signal.

2. *Channel coding*: powerful error correction algorithms associated with the most efficient modulation techniques with regard to the transmission channel characteristics. This step is required to correct transmission errors that have to be near to zero owing to the low redundancy of the signal.

In a way the results have exceeded the original objectives, since it is now possible to transmit many TV programmes (from six to eight or more) in the same RF channel thanks to the efforts of a special group formed in order to obtain a European standard for digital television, and which later became the **DVB** (Digital Video Broadcasting) project office. This chose the MPEG-2 standard as a base for source coding and signal multiplexing, and has defined the channel coding standards for the three main broadcasting media (satellite, cable and terrestrial) as well as a recommended scrambling algorithm.

Sequence of operations

As stated below, many programmes can be transmitted on one RF channel; thus, the sequence of operations to be performed at a transmitter can be roughly illustrated by Fig. 5.2. We will follow the logical order of the functional blocks of this figure when we consider the operations in the following chapters.

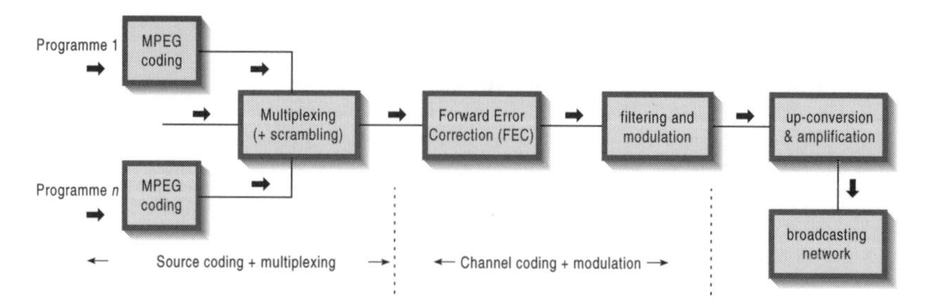

Fig. 5.2 Sequence of main operations at the broadcasting side

5.2 MPEG-2 compression of audio and video signals

We are now going to examine the processes that combine the main steps of the MPEG compression standards for video and audio signals.

5.2.1 Principles of audio and video compression

Video compression makes combined use of many techniques:

- 'general purpose' data compression techniques (**RLC**, **VLC**).
- for fixed pictures (as in **JPEG**) it also exploits the spatial redundancy, e.g. uniform areas within a picture and correlation between neighbouring points (**DCT**) as well as the reduced sensitivity of the eye to small details (thresholding and *quantization*).
- for moving pictures (as in **MPEG**), it also makes use of the temporal redundancy between successive pictures (*movement estimation*).

Run Length Coding (RLC)

When an information source delivers successive message elements that can deliver relatively long series of identical elements, it is advantageous to transmit the code of this element and the number of successive occurrences rather than to repeat the code of the element; this results in a variable compression factor (the longer the series, the bigger the compression factor). This type of coding, which does not lose any information (it is said to be reversible), is commonly used for file compression in computers (ZIP etc.) and is also used in fax machines.

Variable Length Coding (VLC)

VLC coding is based on the fact that the probability of occurrence of an element generated by a source and coded on n bits is not equivalent for all elements amongst the 2^n different possibilities. This means that, in order to reduce the bit-rate required to transmit the sequences generated by the source, it will be advantageous to encode the most frequent elements with less than n bits and the less frequent elements with more bits, resulting in an average length of less than a fixed length of n bits—thus a bit-rate reduction. This is, for instance, the case for the letters of the alphabet in a given language, which allows use of this method for text compression. It is also valid for video images compressed by means of DCT, where energy is concentrated on a relatively small number of coefficients.

The most well-known method for variable length coding is the Huffmann algorithm. This coding is reversible and can be applied to video and audio signals as a complement of other methods (DCT for instance), resulting in very important compression factors.

Discrete Cosine Transform (DCT)

The DCT is a particular case of the Fourier transform which, under certain conditions, decomposes the signal into only one series of harmonic cosine functions that are in phase with the signal. This reduces by half the number of necessary coefficients to describe the signal compared with a Fourier transform.

In the case of a picture, the original signal is a sampled bidimensional signal, and so we will also have a bidimensional DCT (horizontal and vertical directions) which will transform the luminance (or chrominance) discrete values of a block of $N \times N$ pixels into another block (or matrix) of $N \times N$ coefficients representing the amplitude of each of the cosine harmonic functions.

In the transformed block, coefficients on the horizontal axis represent increasing horizontal frequencies, from left to right, and on the vertical axis increasing vertical frequencies from top to bottom.

In order to reduce the complexity of the circuitry and the processing time required, the block size chosen is generally 8×8 pixels (Fig. 5.3), which the DCT transforms into a matrix of 8×8 coefficients (Fig. 5.4) of which the first (top left corner) represents the DC component and is therefore called the DC

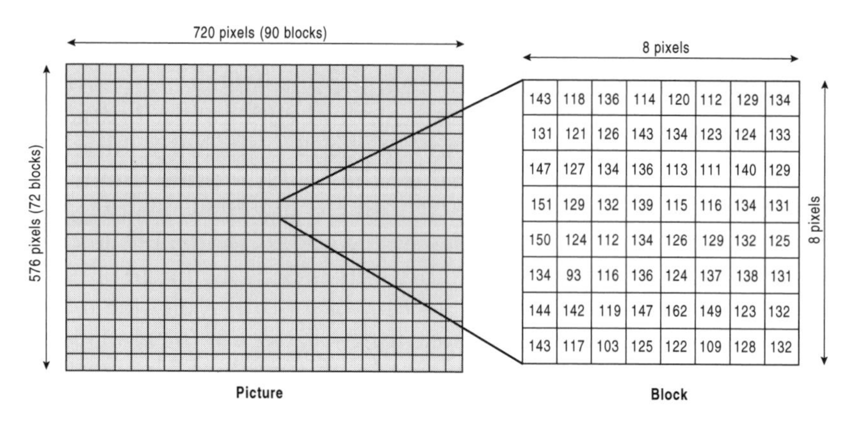

Fig. 5.3 Cutting out blocks of 8 × 8 pixels (values represent the luminance of a pixel)

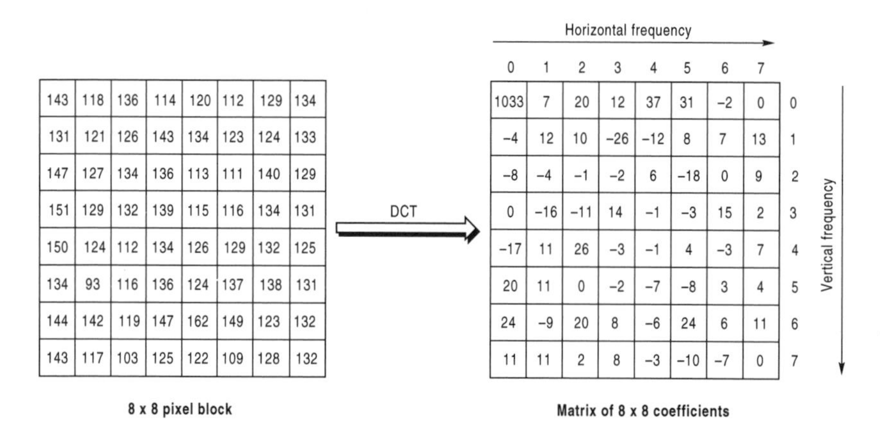

Fig. 5.4 Transformation of a block of 8 × 8 pixels into a matrix of 8 × 8 coefficients using the DCT

coefficient. The bottom right coefficient represents the highest spatial frequency component in the two directions.

The amplitude of the coefficients generally decreases rather quickly with frequency, owing to the smaller energy of high spatial frequencies in most 'natural' images. Consequently, the DCT has the remarkable property of concentrating the energy of the block onto a relatively low number of coefficients, situated in the top left-hand corner of the matrix. In addition, these coefficients are decorrelated from each other. These properties will be profitably used in the next steps of the compression process.

Up to this point, there is no information loss: the DCT transform process is reversible. However, owing to the reduced sensitivity of the eye to high spatial frequencies, it is possible to eliminate (replace by zeros) the values below a certain threshold function of the frequency (an operation known as *thresholding*): this part of the process is obviously not reversible, as some data are thrown away; but it has little effect on the perceived quality.

The remaining coefficients are quantized with an accuracy that decreases with frequency, which again reduces the quantity of information required to encode a block; here again the process is not reversible (Fig. 5.5).

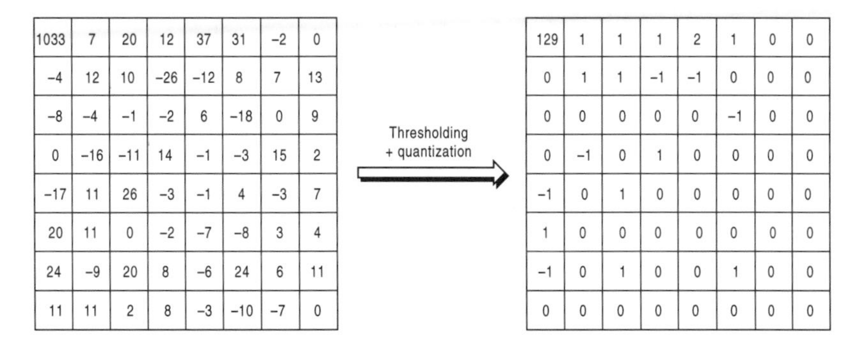

1033	7	20	12	37	31	−2	0
−4	12	10	−26	−12	8	7	13
−8	−4	−1	−2	6	−18	0	9
0	−16	−11	14	−1	−3	15	2
−17	11	26	−3	−1	4	−3	7
20	11	0	−2	−7	−8	3	4
24	−9	20	8	−6	24	6	11
11	11	2	8	−3	−10	−7	0

129	1	1	1	2	1	0	0
0	1	1	−1	−1	0	0	0
0	0	0	0	0	−1	0	0
0	−1	0	1	0	0	0	0
−1	0	1	0	0	0	0	0
1	0	0	0	0	0	0	0
−1	0	1	0	0	1	0	0
0	0	0	0	0	0	0	0

Fig. 5.5 Result of thresholding and quantization

A serial bitstream is obtained by reading the coefficients in zigzag fashion, as shown in Fig. 5.6. This method allows one quickly to obtain a relatively long series of null coefficients, in order to increase the efficiency of the subsequent steps (RLC and VLC).

Compression of fixed pictures (JPEG)

The JPEG norm (Joint Photographic Experts Group) can be considered as a 'toolbox' for fixed picture compression. We will describe the process of its 'lossy' variant, since it has largely inspired the MPEG norms.

It can be described in six main steps.

1. *Decomposition in blocks*: The original in $Y\ C_b\ C_r$ format, is cut up into elementary blocks of 8×8 pixels. Each block is made of 64 numbers ranging from 0 to 255 (when digitized on 8 bits) for luminance and -128 to $+127$ for chrominance C_r and C_b.
2. *Discrete Cosine Transform*: The DCT applied to each $Y\ C_b\ C_r$ block generates, for each one, a new 8×8 matrix (Fig. 5.4) made up of the coefficients of increasing spatial frequencies when moving away from the origin and generally ends with a series of zeros. Therefore, if the block is of uniform luminance

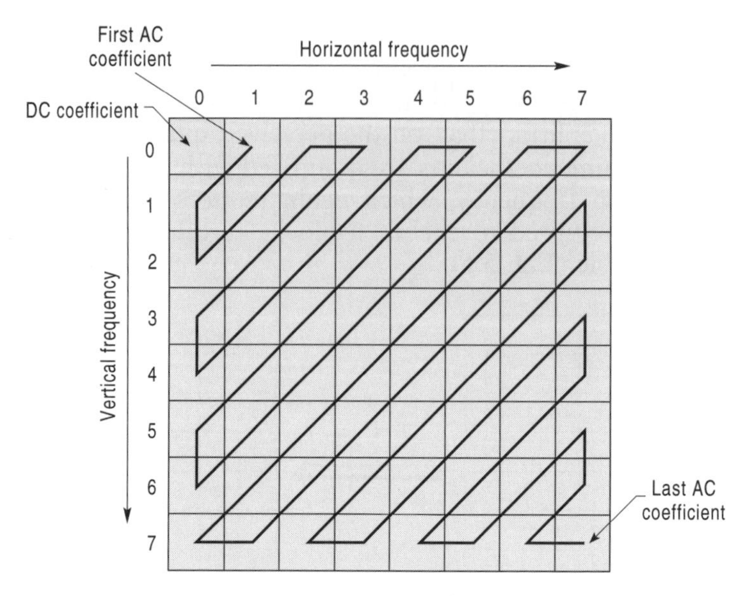

Fig. 5.6 Zigzag reading of the coefficients of the matrix

or chrominance, only the DC coefficient will not be zero, and only this coefficient will have to be transmitted.

3. *Thresholding and quantization*: This step consists of zeroing the coefficients below a predetermined threshold, and quantizing the remaining ones with decreasing accuracy when the frequency increases.
4. *Zigzag scan*: The coefficients are read using a zigzag scan (Fig. 5.6) in order to transform the matrix into a flow of data best suited for the next coding steps (RLC/VLC).
5. *Run Length Coding*: The number of occurrences of zeros is coded, followed by the next non-zero value, which reduces the amount of information to transmit.
6. *Variable Length Coding (Huffman coding)*: This final step uses a conversion table in order to encode the statistically most frequent coefficients with the shortest length.

Only the two last steps (RLC+VLC) ensure a compression factor between 2 and 3.

Compression of moving pictures (MPEG)

In 1990, the ISO created an experts group (**MPEG**, Motion Pictures Experts Group) in order to define a standard for compression of animated pictures with the primary goal of storing them on a CD-ROM with a bit-rate of approximately 1.5 Mb/s (MPEG-1).

In addition to the intrinsic *spatial* redundancy of fixed pictures used in JPEG, coding of animated pictures exploits the important *temporal* redundancy between the successive pictures that make up a video sequence.

Given the very high compression ratio objective (100 compared with 4:2:0), the resolution is reduced from the start by using the SIF format: 360 × 288 @ 25 Hz or 360 × 240 @ 30 Hz. These pictures can be accompanied with stereophonic sound similar to the MUSICAM norm defined for digital radio (**DAB,** Digital Audio Broadcasting).

MPEG-1 video coding (multimedia applications)

The main objective for MPEG-1 was to reach a constant total bit-rate of 1.5 Mb/s, out of which 1.15 Mb/s is for video, with the remaining 350 kb/s being used by audio and auxiliary data. The video coding uses the same principles as a 'lossy' JPEG, to which new techniques that exploit the strong correlation between successive pictures are added, in order to reduce considerably the amount of information required to transmit or store these pictures.

These techniques, known as 'prediction with movement compensation', consist of deducing most of the pictures of a sequence from the preceding and even subsequent pictures with a minimum of additional information representing the differences between pictures. This requires the presence in the MPEG encoder of a movement estimator, which is the most complex component. Fortunately, this part is not required in the decoder.

Since synchronization and random access to a sequence have to be maintained within an acceptable time limit (maximum 0.5 s), the maximum number of pictures that can depend on the same first image is of the order of 10 to 12 for a system operating at 25 pictures/s.

The different MPEG picture types

MPEG defines three types of pictures, which follow each other according to Fig. 5.7.

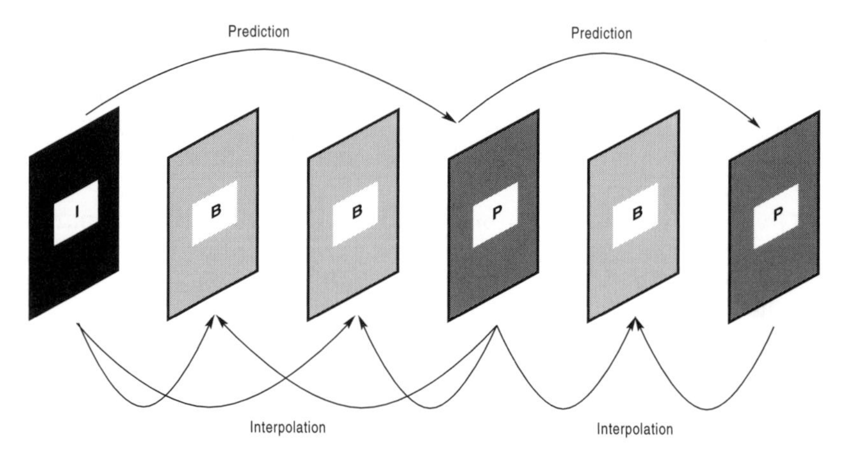

Fig. 5.7 Concatenation of the three types of pictures in MPEG

1. **I (Intra) pictures** are coded without reference to other pictures, in a very similar manner to JPEG, which means that they contain all the information necessary for their reconstruction by the decoder: for this reason, they are the essential entry point for access to a video sequence.
2. **P (Predicted) pictures** are coded from the preceding I or P picture, using the techniques of motion compensated prediction. The compression rate of P pictures is significantly higher than for I pictures.
3. **B (Bidirectional or bidirectionally predicted) pictures** are coded by bidirectional interpolation between the I or P picture which precedes and follows them. B pictures offer the highest compression rate.

Two parameters, M and N, describe the succession of I, P and B pictures (Fig. 5.8):

- M is the distance (in number of pictures) between two successive P pictures,
- N is the distance between two successive I pictures.

The parameters generally used in MPEG-1 are $M = 3$ and $N = 12$.

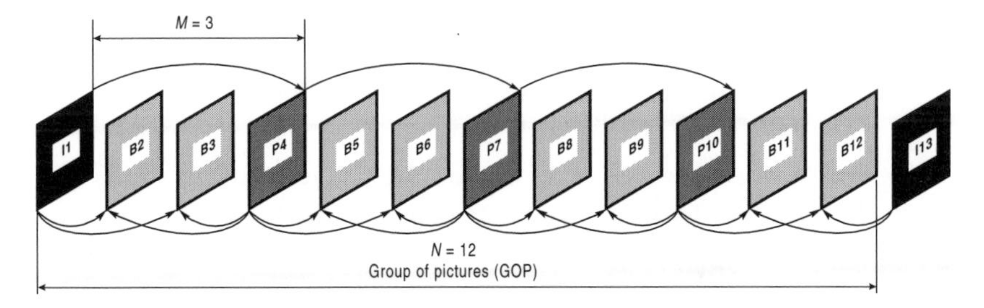

Fig. 5.8 Example of an MPEG group of pictures for $M = 3$ and $N = 12$

Re-ordering of the pictures

At display time, after encoding and decoding, the pictures of the video sequence must obviously be reproduced in the same order as in the original sequence. However, in order to encode or decode a B (Bidirectional) picture, the encoder and the decoder will need the preceding I (Intra) or P (Predicted) picture as well as the subsequent P or I picture. Consequently, the order of the pictures will be modified before coding in order that the encoder and the decoder have at their disposal the I and/or P pictures required for the processing. This reordering takes place in the decoder at display time.

Decomposition of an MPEG video sequence in layers

MPEG defines a hierarchy of *layers* within a video sequence, as illustrated by Fig. 5.9.

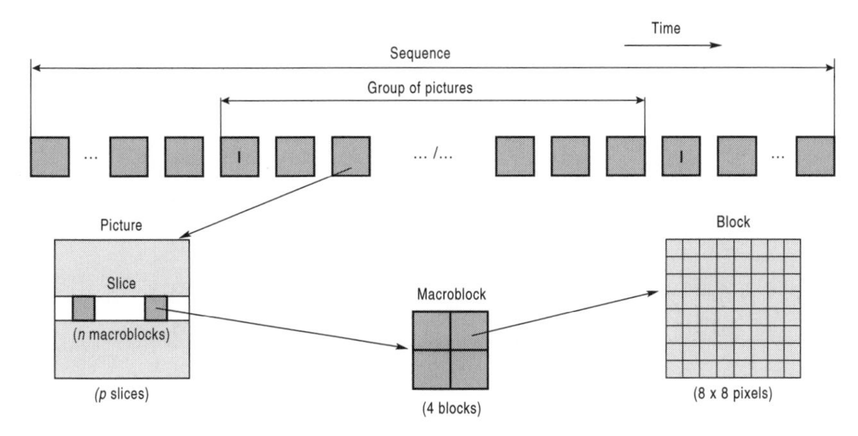

Fig 5.9 Hierarchy of MPEG video layers

Starting from the top level, the successive layers are as follows.

- *Sequence*: defines the context for the sequence (video parameters, scanning standard etc.).
- *Group of pictures (GOP)*: this is the layer determining the random access to the sequence, which always starts with an I picture. In the above example ($M = 3$, $N = 12$), the GOP is made of 12 pictures (Fig. 5.8).
- *Picture*: this is the elementary display unit; it can be of I, P or B type.
- *Slice*: this is defined as a suite of contiguous macroblocks, most often a complete row of macroblocks.
- *Macroblock*: this is the layer used for movement estimation/compensation. A macroblock has a size of 16 × 16 pixels

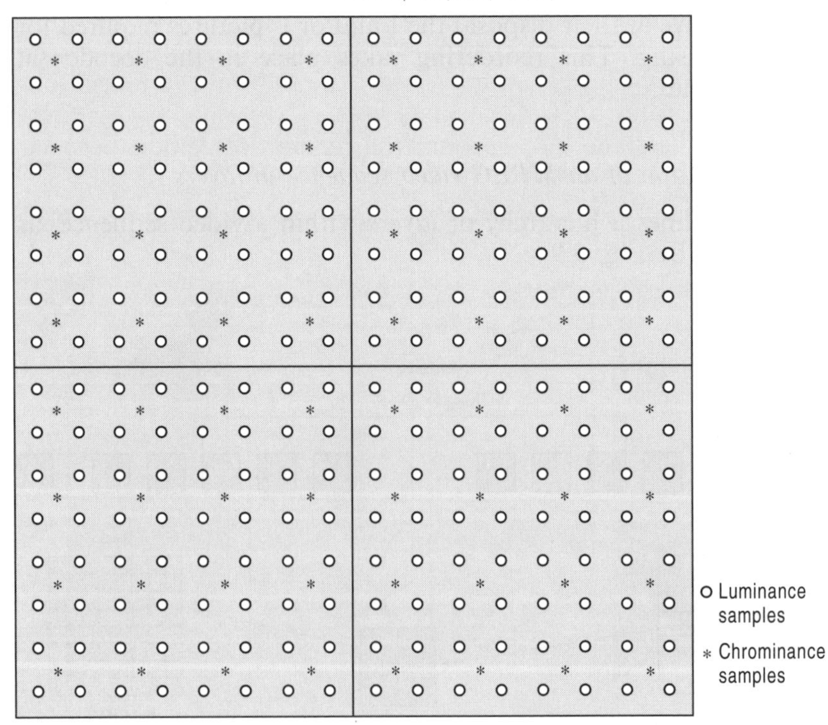

Fig. 5.10 Composition of a 4:2:0 macroblock (○ = Y samples, * = C_b and C_r samples)

(four blocks of luminance and two blocks of chrominance, one C_r and one C_b, of 8 × 8 pixels each) (Fig. 5.10).

- *Block*: as in JPEG, a picture is divided into blocks of 8 × 8 pixels. The block is the layer where the DCT takes place.

Motion estimation and compensation

In a sequence of moving pictures, moving objects lead to differences between corresponding zones of consecutive pictures; so there is no obvious correlation between these two zones. *Motion estimation* consists of finding a motion vector which ensures the correlation between an 'arrival' zone on the second picture and a 'departure' zone on the first picture. This is done at the macroblock level (16 × 16 pixels) by moving a macroblock of the current picture within a search window from the previous picture, and comparing it with all possible macroblocks of that window in order to find the one that most resembles it (*block matching*). The difference in position of the two matching macroblocks gives a motion vector (Fig. 5.11) which will be applied to all three components of the macroblock (Y, C_b, C_r). Only the macroblocks that differ from the picture(s) used for prediction will need to be encoded—this will substantially reduce the amount of information.

For B pictures, the intermediate vectors are obtained by interpolations that take into account the temporal position between

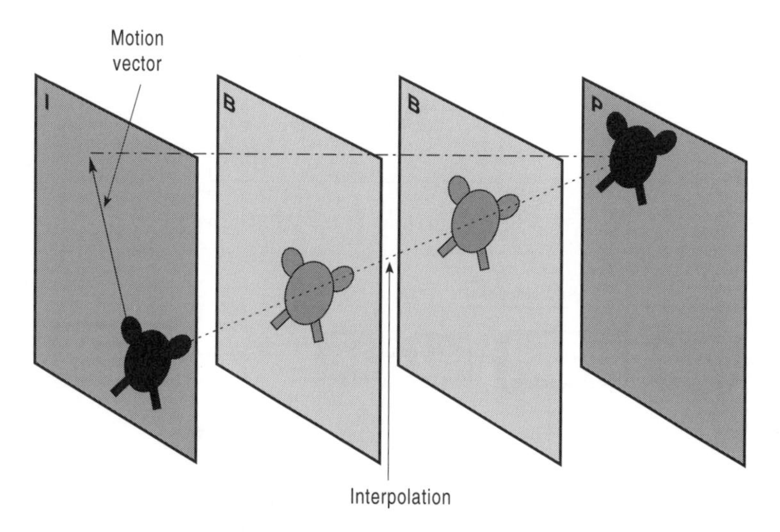

Fig. 5.11 Simplified illustration of motion compensation

the I and/or P pictures. The difference between the block to be encoded and the one that most resembles it is found by the motion estimator, and the difference is encoded following a coding similar to the one used for coding the blocks of the I pictures (DCT, quantization, VLC).

Output bit-rate control

The bit-rate at the output of the encoder must generally be kept constant; for this purpose, a buffer memory (FIFO) is used—the 'filling' condition of which is monitored and maintained within predefined limits by acting on the resolution of the quantization coefficients. Figure 5.12 shows a very schematic block diagram of the MPEG encoder (however, it does not give a good idea of its real complexity).

The decoder does not have to perform movement estimation

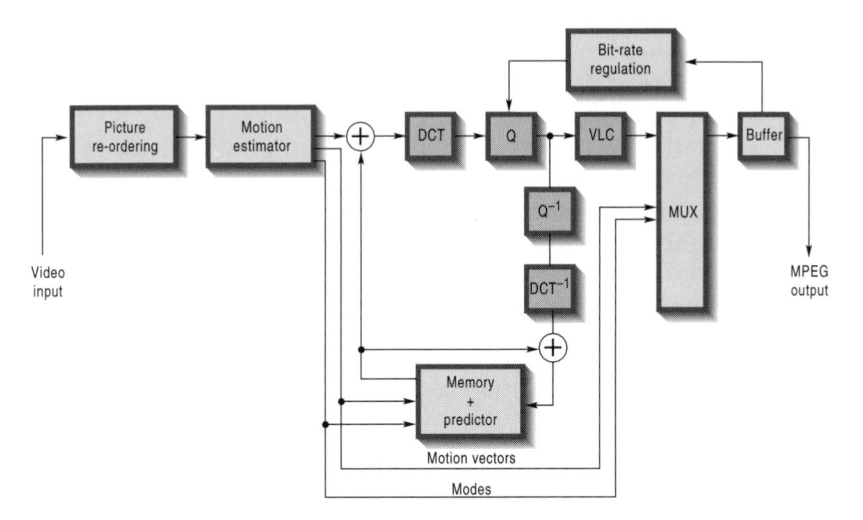

Fig. 5.12 Schematic diagram of the MPEG encoder

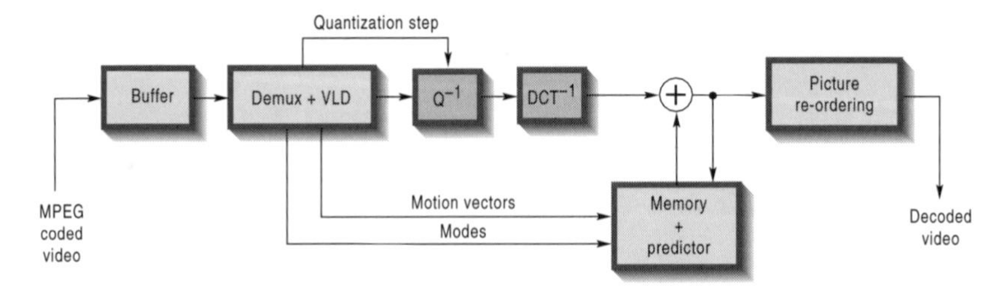

Fig. 5.13 Schematic diagram of the MPEG decoder

and so is much simpler (Fig. 5.13), which was one of the main objectives of the standard.

MPEG-2 video coding

MPEG-1 video coding does not offer sufficient resolution for broadcasting applications and, in addition, it is not compatible with interlaced pictures. Hence, the DVB group, which established the European digital TV standard, chose MPEG-2 as a source coding standard for European digital TV transmissions.

There is an ascending compatibility between MPEG-1 and MPEG-2, which means that an MPEG-2 decoder will be able to decode MPEG-1 elementary streams.

MPEG-2 profiles and levels

The MPEG-2 standard defines five *profiles*, which determine the set of compression tools used, and four *levels*, which define the resolution of the picture, ranging from SIF to HDTV.

Figure 5.14 illustrates the main characteristics of the different levels and profiles of MPEG-2 (source: DVB Project Office).

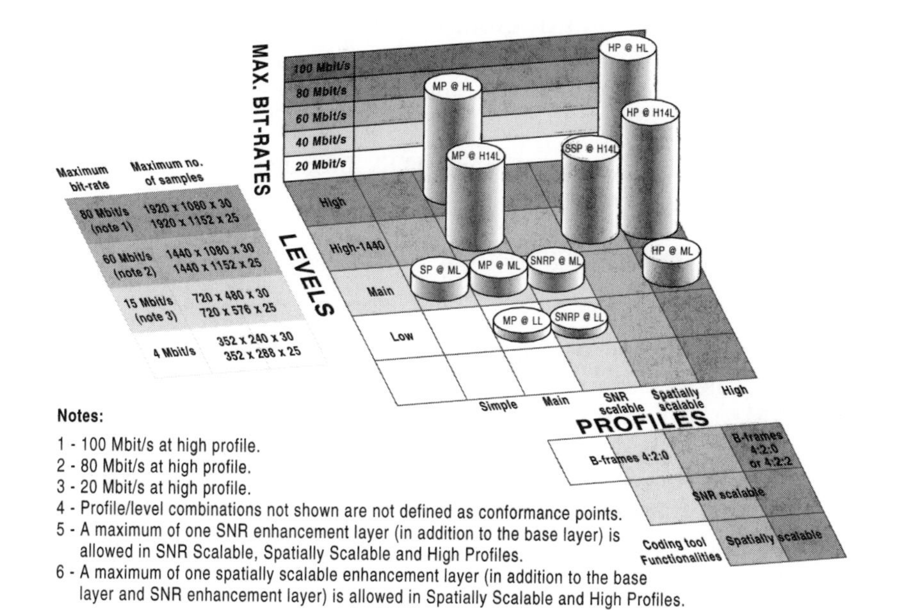

Notes:

1 - 100 Mbit/s at high profile.
2 - 80 Mbit/s at high profile.
3 - 20 Mbit/s at high profile.
4 - Profile/level combinations not shown are not defined as conformance points.
5 - A maximum of one SNR enhancement layer (in addition to the base layer) is allowed in SNR Scalable, Spatially Scalable and High Profiles.
6 - A maximum of one spatially scalable enhancement layer (in addition to the base layer and SNR enhancement layer) is allowed in Spatially Scalable and High Profiles.

Fig. 5.14 MPEG-2 levels and profiles (source: *Going Ahead with Digital Television*; © DVB Project Office 1995)

The four levels can be described as follows:

- the *low* level corresponds to the SIF resolution used in MPEG-1 (up to 360 × 288)
- the *main* level corresponds to standard 4:2:0 resolution (up to 720 × 576)
- the *high-1440* level is aimed at HDTV (resolution up to 1440 × 1152)
- the *high* level is optimized for wide-screen HDTV (resolution up to 1920 × 1152)

In the case of profiles, it is a little more complex.

- The *simple* profile is defined in order to simplify the encoder and the decoder, at the expense of a higher bit-rate, as it does not use bidirectional prediction (B pictures).
- The *main* profile is today the best compromise between compression rate and cost, as it uses all three image types (I, P, B) but leads to more complex encoders and decoders.
- The *scalable* profiles (hierarchy coding) are intended for future use, for instance to transmit (in a compatible way) the same programme in standard resolution and in high definition (HD) or, alternatively, to allow a basic quality reception in the case of difficult receiving conditions and an enhanced quality in good receiving conditions.
- The *high* profile is intended for HDTV broadcast applications in 4:2:0 or 4:2:2 format.

There is an ascending compatibility between profiles, and a decoder of a given profile will be able to decode all lower profiles (see the left part of Fig. 5.14).

The combination retained for short-term consumer broadcast applications in Europe, is known as *Main Profile at Main Level* (MP@ML). It is based on interlaced pictures in 4:2:0 format with a resolution of up to 720 × 480 @ 30Hz or 720 × 576 @ 25 Hz with a toolbox including coding of I, P and B pictures. Depending on the compromise struck between bit-rate and picture quality, and the nature of the pictures to be transmitted, the bit-rate will generally be between 4 Mb/s (giving a quality similar to PAL or SECAM) and 9 Mb/s (near CCIR-601 studio quality).

Apart from the resolution of the original picture and the processing of interlaced pictures, all the processing described for

MPEG-1 is valid for MPEG-2 (MP@ML) encoding and decoding and, in particular, the layer hierarchy (from block to sequence).

For the best results, interlaced pictures will have to be processed in different ways depending on the importance of movements between the two fields of a picture: the extreme cases are, on the one hand, pictures originating from cinema films, where the two fields come from the same cinema picture (at least in 50 Hz systems) and, on the other hand, TV pictures from sporting events, where differences due to motion between the two fields of a picture can be important.

For the coding of interlaced pictures, MPEG-2 permits a choice between two image structures called a *frame* and a *field*:

- The frame structure (also called progressive) is best suited for the cases where there is little movement between two successive fields. Macroblocks and blocks are then cut out of the complete frame (Fig 5.15).
- The field structure (also called interlaced) is preferable when there are important movements between successive fields; in this

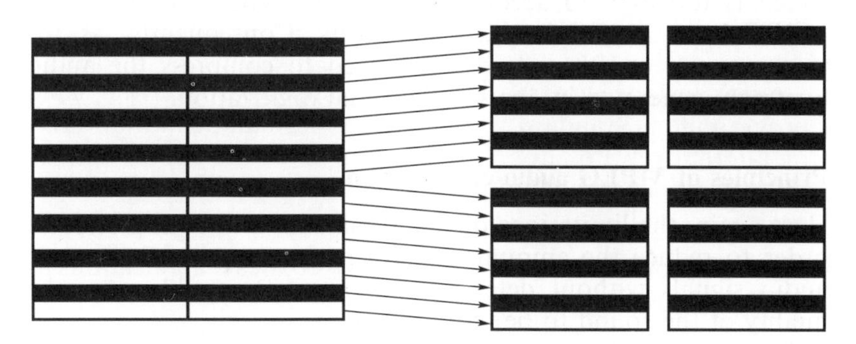

Fig. 5.15 Cutting blocks out of macroblocks (frame mode)

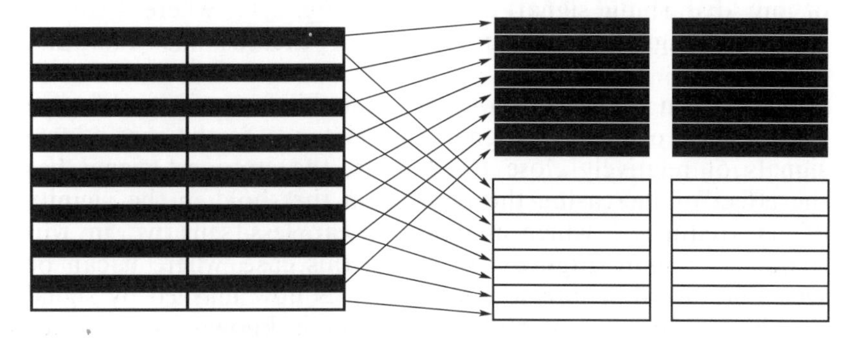

Fig. 5.16 Cutting blocks out of macroblocks (field mode)

case, macroblocks are cut out of one field only (Fig. 5.16), which is then considered as an independent picture.

5.2.2 Compression of audio signals

The use of digital audio has become widespread since the introduction of the Compact Disc at the beginning of the 1980s. In order to obtain hi-fi quality, a sampling scheme at 44.1 kHz with a resolution of 16 bits has been chosen in order to guarantee a bandwidth of 20 kHz and a high signal-to-noise ratio as well as a wide dynamic range (96 dB in theory).

The net result is a bit-rate of $44.1 \times 16 \times 2 = 1411.2$ kb/s for a stereophonic signal. Two other sampling frequencies are commonly used: 32 kHz for NICAM and D2MAC, and 48 kHz for professional audio recording as well as DAT, MiniDisc and DCC. MPEG audio foresees the possibility of using, for the source signal, any of the above three sampling frequencies. The corresponding bit-rates will vary accordingly. However, they will stay in the same order of magnitude (between 1.0 and 1.5 Mb/s), which is too high to accompany a picture coded according to MPEG-1 or MPEG-2 video standards. Consequently, as for video, one is confronted with the need to compress the audio bitstream generated by the simple signal digitization.

Principles of MPEG audio compression

Here again, the limitations of the human ear will be exploited in order to reduce the amount of information required to encode audio signals without deteriorating, in a perceptible way, the quality of the sound to be reproduced.

The ear sensitivity curve, which represents the audibility or perception threshold as a function of frequency (in the absence of any ‘disturbing’ signal) is shown in Fig. 5.17, where it can be seen that signal A is audible, as it exceeds the audibility threshold.

It has been suggested, however, that this curve is modified in the presence of multiple signals: for instance, in the case of two signals of relatively close frequencies, the strongest signal has the effect of increasing the perception threshold in the vicinity of its frequency, which makes the ear less sensitive in this frequency region. Figure 5.18 shows this case, where it can be seen that signal A, previously audible, is now masked by signal B, which is more powerful. This effect is known as *frequency masking*.

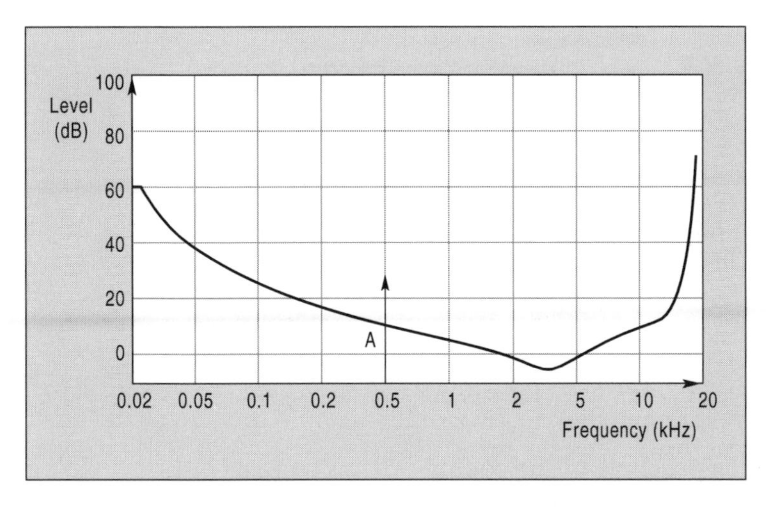

Fig. 5.17 Sensitivity of the ear as a function of frequency (signal A is audible)

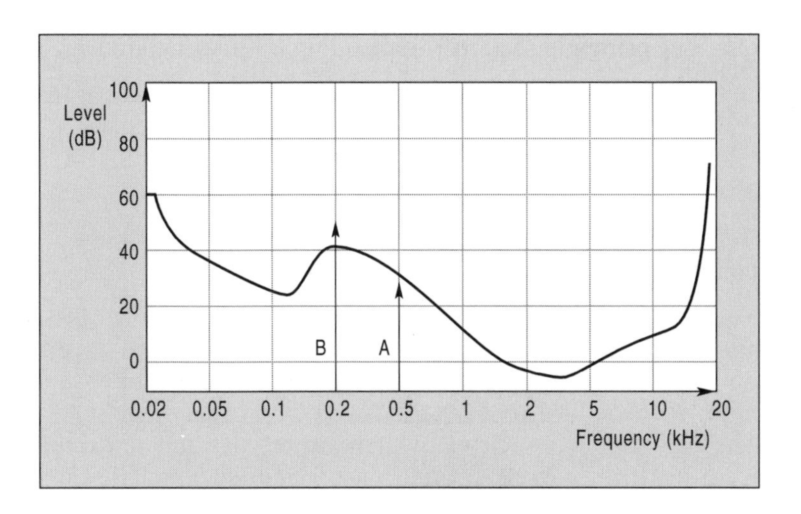

Fig. 5.18 Frequency masking effect (signal A is masked by signal B)

There is also another effect, called *temporal masking*. Here, a sound of strong amplitude also masks sounds immediately preceding it or following it in time, as illustrated by Fig. 5.19.

Experiments have been conducted that have led to the definition of a *psycho-acoustical* model of human hearing. This model is the basis of the concept of a *perceptual encoder*, which is characterized by a masking curve and a quantization variable that is a function of the signals to be encoded.

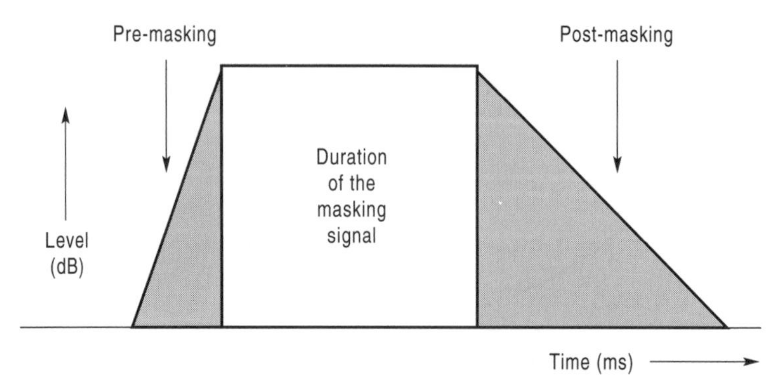

Fig. 5.19 Temporal masking effect

The principle of the coding process consists of first dividing the audio frequency band into 32 sub-bands of equal width by means of a *polyphase* filter bank. The output signal from a sub-band filter corresponding to a duration of 32 **PCM** samples is called a *sub-band sample.* The principle of perceptual coding is illustrated in Fig. 5.20.

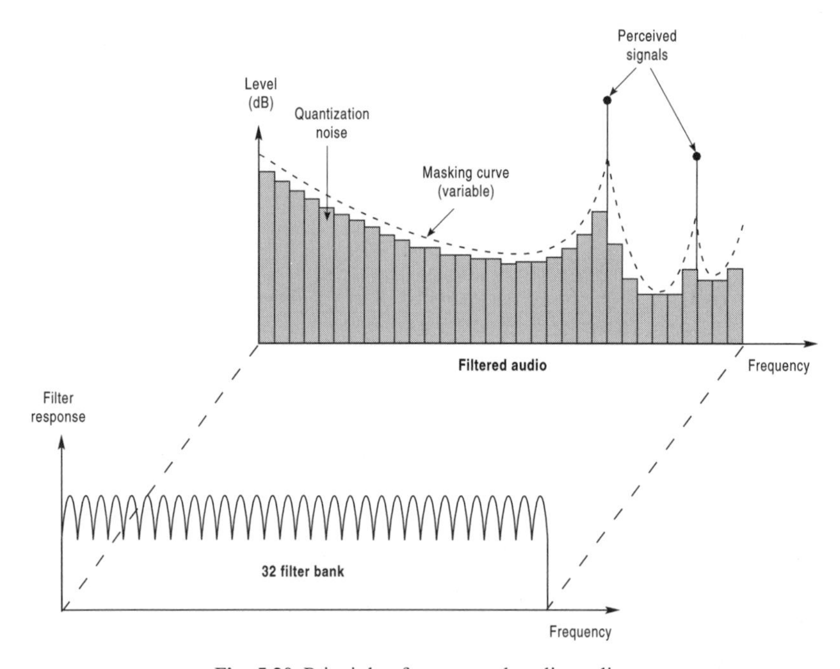

Fig. 5.20 Principle of perceptual audio coding

The psycho-acoustical model allows elimination of all sub-band signals below their perception threshold, as they would not be heard by the listener. The model also defines the required quantization accuracy for each sub-band in order that the quantization noise stays below the audibility threshold for this sub-band.

In this way, frequency regions where the ear is more sensitive, can be quantified with greater accuracy than other regions. Simplified block-diagrams of the MPEG audio encoder and decoder are shown in Figs 5.21 and 5.22 respectively.

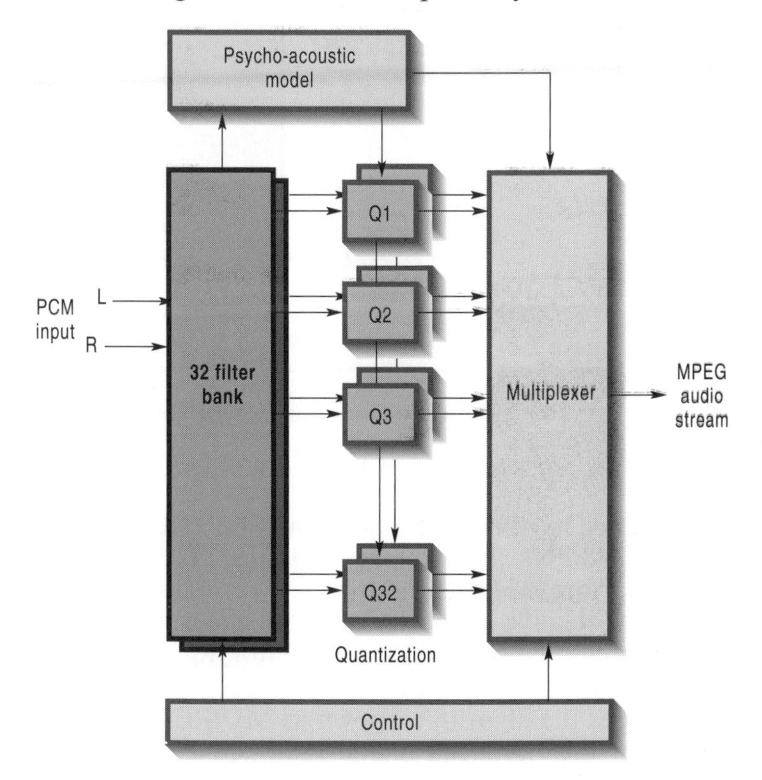

Fig. 5.21 Principle behind the MPEG audio encoder

Analysis of the signal in order to determine the masking curve and quantization accuracy is not carried out for each PCM sample, but instead is performed in a time interval called a *frame.*

All information necessary for sound decoding is supplied at the frame level, which is the smallest unit for random access to the sequence (comparable to the group of pictures for video).

The layers of MPEG-1 audio coding

The MPEG audio standard defines three coding *layers* which offer very different compression rates for a given perceived audio quality.

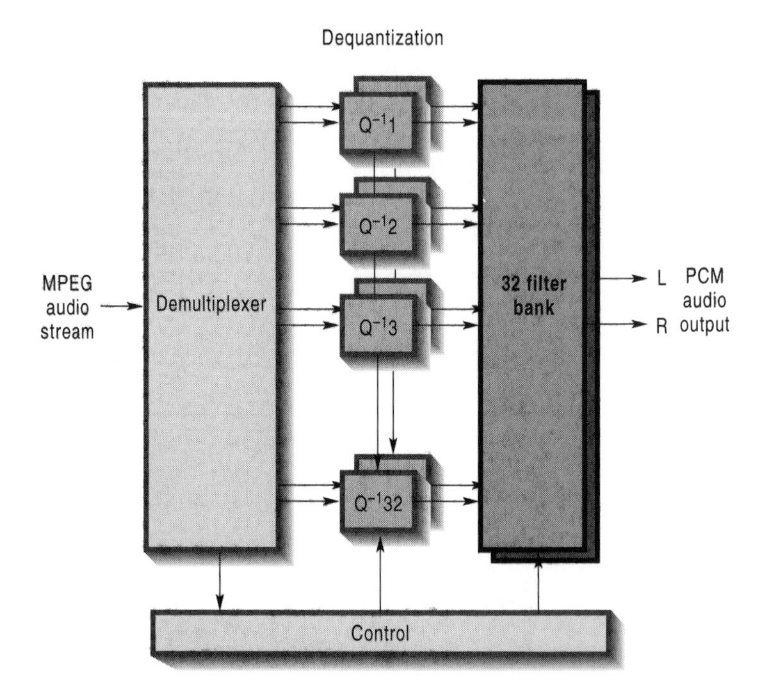

Fig. 5.22 Principle behind the MPEG audio decoder

Layer I uses one fixed bit-rate among the 14 possible which range from 32 to 448 kb/s. Hi-fi quality requires 192 kb/s per audio channel, therefore 384 kb/s in stereo. The main advantage of Layer I is the relative simplicity of the encoder and decoder. It has mainly been used by Philips, its inventor, for its digital audio cassette (DCC).

Layer II uses the algorithm known as MUSICAM, which was developed for the European digital radio standards (DAB) and television (DVB). For an equivalent audio quality, Layer II requires a 30 to 50% smaller bit-rate than Layer I, at the expense of a moderate increase in complexity for the encoder and the decoder. The bit-rate is fixed and chosen from 32 to 192 kb/s per channel, the hi-fi quality being obtained from 128 kb/s per channel (256 kb/s in stereo).

Layer III allows a variable bit-rate and a compression rate approximately twice as high as Layer II, but the encoder and decoder are substantially more complex. In addition, the encoding/decoding time is much longer. Hi-fi quality requires only 64 kb/s per channel (128 kb/s for stereo). Layer III is mainly intended for unidirectional applications using low bit-rate media (ISDN for instance).

As for MPEG-2 video levels and profiles, MPEG audio layers are upwards compatible, which means that a Layer III decoder will be able to decode Layers I and II and a Layer II decoder will be able to decode Layer I.

For audio coding, the DVB digital TV standard uses layer II of the MPEG-1 audio specification (MPEG-1 layer II), which foresees four main audio modes:

- *stereo*: the left and right channels are coded completely independently
- *joint_stereo*: exploits the redundancy between the left and right channels to reduce the bit-rate
- *dual_channel*: two independent audio channels (bilingual for instance)
- *mono*: one audio channel only.

Multichannel extensions

The MPEG-1 audio Layer II specification does not foresee direct multichannel (surround) sound, which is, however, possible through the same methods as in analogue TV, such as Dolby ProLogic, i.e. by means of a decoder integrated in the TV set or external to it. Two standards are in competition for an optimized multichannel audio system (for DVD video disc): MPEG-2 audio, which is compatible with a stereo MPEG-1 Layer II decoder, and is favoured by Europe, and Dolby AC3, which requires a specific decoder and is the choice of the USA.

5.3 Signal multiplexing, scrambling and conditional access

5.3.1 Organization of the MPEG-2 multiplex (programme and transport streams)

An MPEG-2 encoding system has to ensure video and audio data encoding for each programme, it must add the information necessary for synchronization and the resources required for decoding, and then multiplex the information coming from all programmes with other 'private' data.

The elementary streams delivered by the audio and video encoders are put into 'packets' which form the audio, video and private 'Packetized Elementary Streams' (**PESs**). The PESs start with a packet header, the format of which is shown in Fig. 5.23.

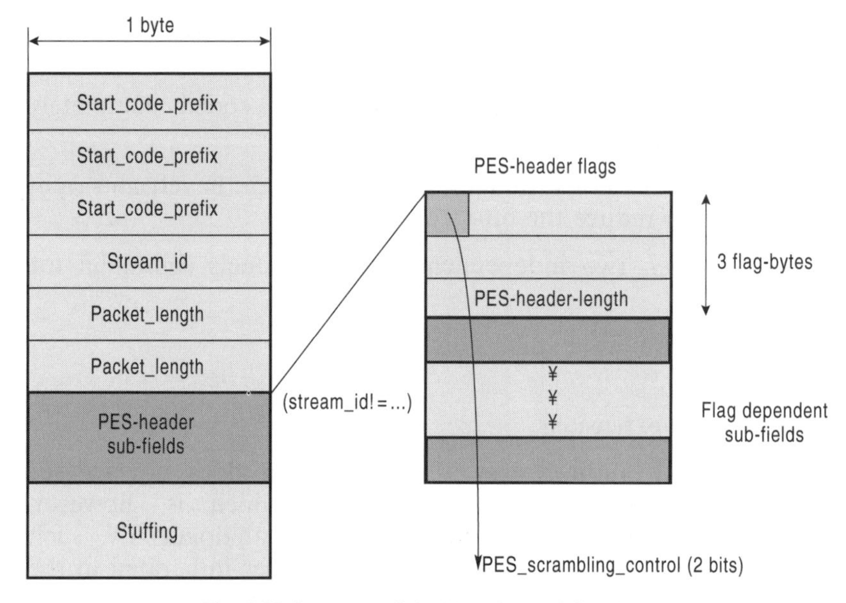

Fig. 5.23 Structure of the MPEG-2 PES header

The MPEG-2 standards allow two different methods of PES multiplexing, depending on the application, as shown in Fig. 5.24.

1. Programme stream

The MPEG-2 programme stream is made of one or more PESs (video, audio, private) which must necessarily share the same System Time Clock (**STC**). This type of stream is suited for applications where the transmission channel or storage medium is supposed to introduce only a very small number of errors (a *quasi-error-free medium*), as is in multimedia applications based on CD-ROM or hard disk. In these cases, packets can be relatively long (2048 bytes for instance). This kind of multiplex is used for storage of MPEG-2 video on Digital Versatile Disk (**DVD**).

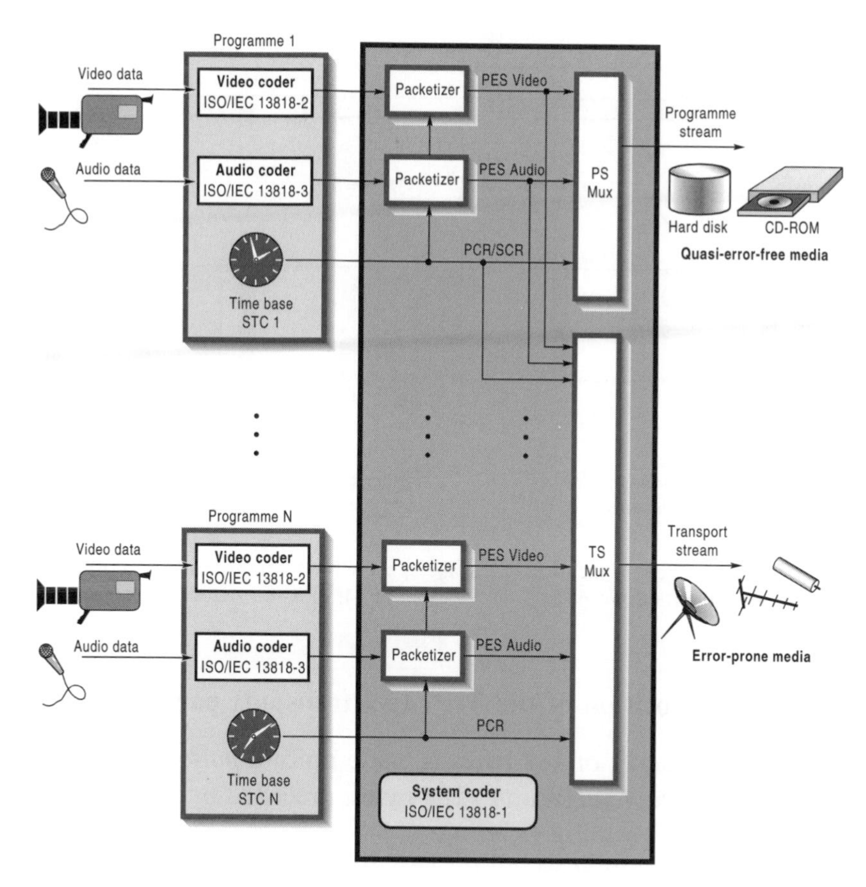

Fig. 5.24 Conceptual diagram of the generation of MPEG-2 transport and programme streams

2. Transport stream

The MPEG-2 transport stream is primarily intended for transport of TV programmes over long distances via transmission supports or in environments susceptible to the introduction of relatively high error rates (an *error-prone medium*). In these cases, the packet length should be relatively short in order to allow implementation of efficient correction algorithms, which will be detailed in section 5.4.

The length of the MPEG-2 *transport packet* therefore has been fixed at 188 bytes for transmission of TV programmes by satellite, cable or terrestrial, following the European DVB standard. Fig. 5.25 illustrates the process for the creation of an MPEG-2 transport stream.

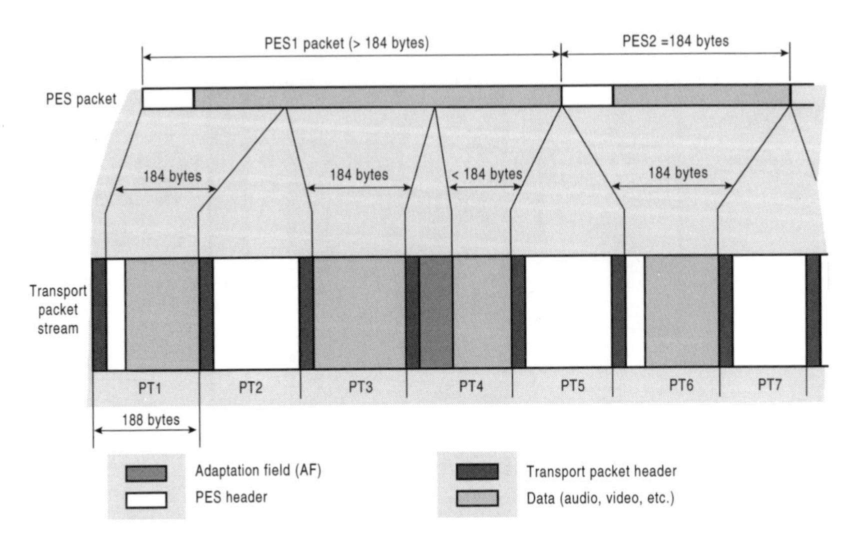

Fig. 5.25 Arrangement of the PESs that make up an MPEG-2 transport stream (PES1 is split between the transport packets PT1, PT3 and PT4; PES2 fits exactly into transport packet PT6)

5.3.2 Composition of the MPEG-2 transport packet

A transport packet of 188 bytes is made up of a *packet header* of 4 bytes and a *payload* of up to 184 bytes, preceded by an optional *adaptation field*, see Fig. 5.26.

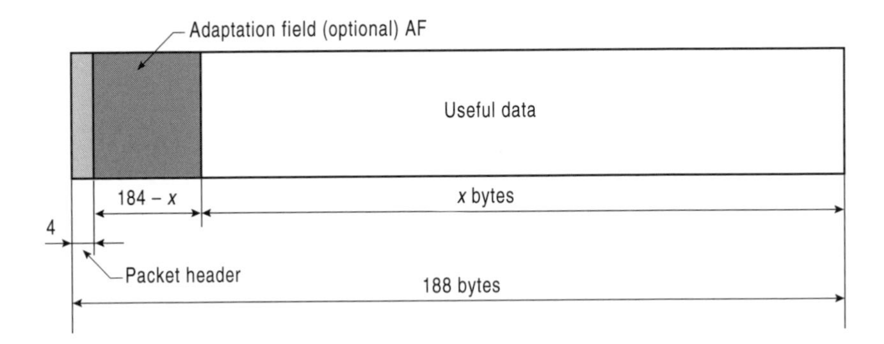

Fig. 5.26 Composition of the transport packet

In this context, the payload means the data from the PES composing the TV programmes to which are added a certain number of data, allowing the decoder to find its way in the MPEG-2 transport stream. Each transport packet is identified by a **PID**

(Packet IDentifier) contained in the packet header (Fig. 5.27), which starts with a synchronization byte (value 47h).

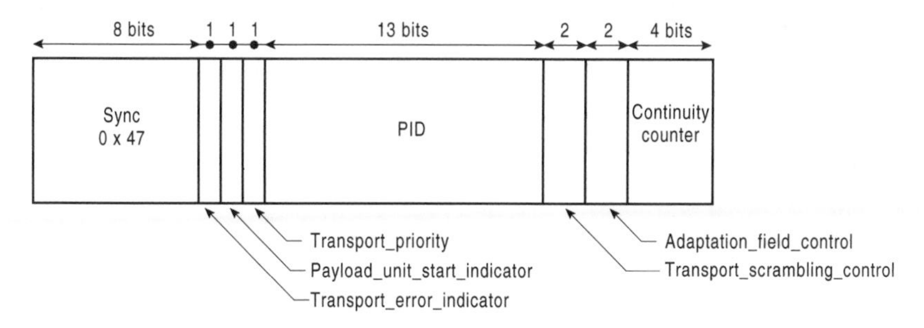

Fig. 5.27 Detail of the transport packet header

A given transport packet can only carry data coming from one PES packet, and a PES packet always starts at the beginning of the payload part of a transport packet and always ends at the end of a transport packet (Fig. 5.25).

As transport packets (188 bytes) are generally (much) shorter than PES packets (e.g. 2048 bytes), PES packets will have to be divided into data blocks of 184 bytes. The length of PES packets is generally not an exact multiple of 184 bytes, so the last transport packet carrying a PES packet will have to start with an *adaptation field* (AF), the length of which will be equal to 184 bytes less the number of bytes remaining in the PES packet.

In addition to this 'stuffing' function, the adaptation field is also used to carry the reference clock of the programme (Program Clock Reference, **PCR**).

5.3.3 Organization of the transport multiplex: MPEG-2 tables

The MPEG-2 transport multiplex generally carries many programmes, each composed of one or more PESs (Program Elementary Streams). In order that the decoder can find its way in these streams, MPEG-2 has specified *tables* known as **PSI** (Program Specific Information) which give information on the contents and the organization of the transport multiplex. These tables are cut in sections of 1024 bytes.

Program Association Table (PAT)

The presence of this table is mandatory, and it is carried by the packets of PID = 0. Its purpose is to indicate, for each programme carried by the transport multiplex, the link between the programme number (from 0 to 65 535) and the PID of packets carrying a 'map' of the programme (Program Map Table, PMT). The PAT is always broadcast 'in the clear' even if all programmes of the multiplex are scrambled.

Program Map Table (PMT)

There is one PMT for each programme present in the multiplex. It indicates mainly (in the clear) the PID of the elementary streams constituting the programme and, optionally, other private information relating to the programme, which can eventually be scrambled (for example the ECM, the first piece of information necessary to unscramble programmes with conditional access, see below). The PMT can be transported by packets of arbitrary PID, defined by the broadcaster and indicated in the PAT (except 0 and 1).

Conditional Access Table (CAT)

This table must be present if at least one programme in the multiplex has conditional access. It is transported by the packets of PID = 1 and indicates the PID of packets carrying the EMM for one (or more) conditional access systems (the second piece of information necessary for unscrambling programmes with conditional access, see below).

Private Tables

These tables carry private data, which are either in free format or in a format similar to the CAT, except for the section length, which can be up to 4096 bytes.

MPEG-2 allows supplementary Service Information (**SI**), which has been grouped under the DVB-SI description; these tables allow the automatic configuration of the receiver and makes user 'navigation' easier in the numerous services available. The DVB-SI information is made up of four main tables and three optional tables.

Main tables of DVB-SI

Network Information Table (NIT)

This table, as its name implies, carries information specific to a network made up of more than one RF channel (i.e. more than one transport stream), for example frequencies or channel numbers used by the network, which the receiver can use to configure itself.

This table is, by definition, the programme number 0 of the multiplex.

Service Description Table (SDT)

This table lists the names and other parameters associated with each service in the multiplex.

Event Information Table (EIT)

This table is used to transmit information relative to events occurring or going to occur in the current transport multiplex or even other transport streams.

Time and Date Table (TDT)

This table is used to update the internal real time clock of the set-top box.

Optional tables of DVB-SI

Bouquet Association Table (BAT)

This table can be used as a tool for grouping services that the set-top box may use to present the various services to the user, for example by way of the EPG. A given service or programme can be part of more than one bouquet.

Running Status Table (RST)

This table is transmitted only once for a quick update of the status of one or more events at the time when this status changes, and not repeatedly as with the other tables.

Stuffing Tables (ST)

These tables are used, for instance, to replace previously used tables which have become invalid.

The repetition frequency of these tables is not rigidly imposed by the standard, but it must be high enough (10 to 50 Hz) to allow the decoder to access the desired programme quickly enough.

The part of these tables visible to the user is the interactive presentation of the Electronic Programme Guide (**EPG**) generally associated with the network or 'bouquet' by means of the information carried by the DVB-SI tables. It allows easy navigation through the numerous programmes and services at the user's disposal.

5.3.4 Scrambling and conditional access

The share of free access programmes among analogue TV transmissions by cable or satellite is continuously decreasing and, following this trend, most digital transmissions today are 'pay-TV' programmes. Billing forms can be much more diversified (conventional subscription, pay-per-view, near video on demand and so on), which is made easier by the high available bit-rate of the system and the presence of a 'return channel' today provided by a modem.

The DVB standard, as explained in the previous chapter, foresees the transmission of access control data carried by the CAT and other private data packets indicated by the PMT. The standard also defines a Common Scrambling Algorithm (**CSA**) used by most (but not all) European service providers. The Conditional Access system (**CA**) itself is not defined by the DVB standard, as most operators did not want a common system, both for purely commercial reasons (subscriber database management) and security reasons (piracy). However, in order to avoid the subscriber who wishes to access networks using different conditional access systems having a pile of boxes (one set-top box per network), the DVB standard envisages two options.

1. *Simulcrypt*: this process allows access to the same programme by means of more than one conditional access system; it requires an agreement between operators. In this case, the transport multiplex will have to carry the conditional access packets for each of the supported systems, which must use the same scrambling algorithm (the CSA of the DVB).

2. *Multicrypt*: all the functions required for conditional access and descrambling are contained in a *detachable* module in a PC-CARD (ex PCMCIA) form factor which is inserted into the transport stream data path. This is done by means of a standardized interface (Common Interface, **DVB-CI**), which also includes the processor bus for information exchange between the module and the set-top box. The set-top box can have more than one DVB-CI slot, thus allowing connection of many conditional access modules. For each different conditional access and/or scrambling system required, the user can connect a module which, in addition to the conditional access software, can contain a smart card interface and a suitable descrambler. The DVB-CI connector may also be used for other purposes (for example data transfers).

Principles of the scrambling system in the DVB standard

Given the very delicate nature of this part of the standard, it is understandable that only very general details are available, implementation details are only accessible to network operators and equipment manufacturers under a non-disclosure agreement. In order to resist attacks from hackers for as long as possible the scrambling algorithm is based on ciphering with two layers, each palliating the eventual weaknesses of the other:

- a *block* layer using blocks of eight bytes
- a *stream* layer (pseudo-random byte generator)

The scrambling algorithm uses two control words (even and odd) alternated with a frequency of the order of 2 s in order to make the pirate's task more difficult. One of the two encrypted control words is transmitted in the Entitlement Control Messages (ECMs) during the time the other is in use, so that the control words have to be stored temporarily in the registers of the descrambling device. There is also a *default* control word (which could be used for free access scrambled transmissions).

Conditional access mechanisms

The information required for descrambling is transmitted in specific Conditional Access Messages (CAMs), which are of two types: Entitlement Control Messages (ECMs) and Entitlement Management Messages (EMMs).

These messages are generated from three different types of input data:

- a *control_word*, which is used to initialize the descrambling sequence
- a *service_key* used to scramble the control word for a group of one or more users
- a *user_key* used for scrambling the service key.

The process for generating ECM and EMM is illustrated in Fig. 5.28.

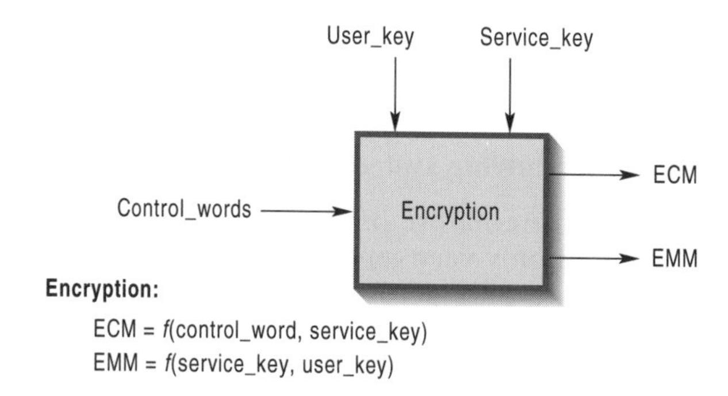

Fig. 5.28 Schematic illustration of the ECM and EMM generation process

- ECMs (Entitlement Control Messages) are a function of the *control_word* and *service_key* and are transmitted approximately every 2 s.
- EMMs (Entitlement Management Messages) are a function of the *service_key* and *user_key* and are transmitted approximately every 10 s.

At reception, the principle of decryption consists of recovering the service_key from the EMM and the user_key contained, for instance, in a smart card. The service_key is then used to decrypt ECM in order to find the control_word which allows initialization of the descrambling device. Figure 5.29 schematically illustrates this process.

Figure 5.30 illustrates the process followed to find the ECM and EMM required to descramble a given programme (here programme number 3).

The Programme Association Table (PAT), rebuilt from sections in packets with PID = 0, indicates the PID (M) of the

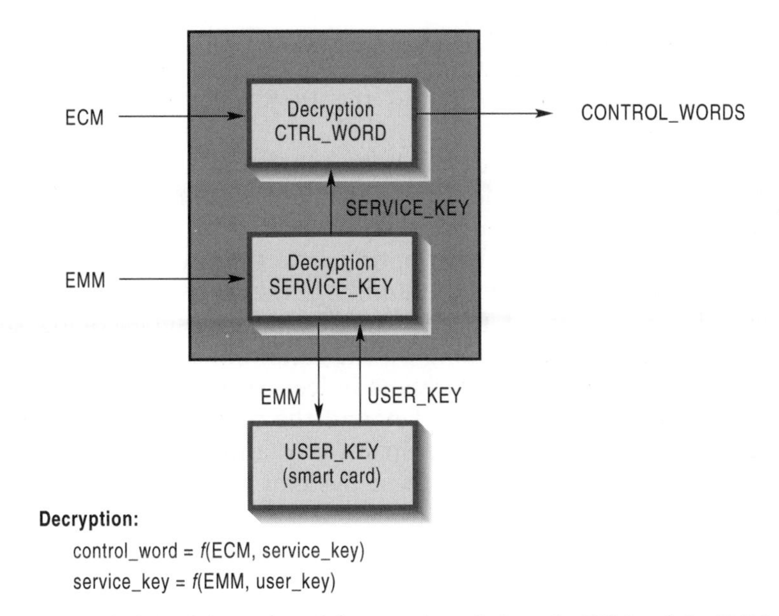

Fig. 5.29 Principle of decryption of the control words from the ECM and the EMM

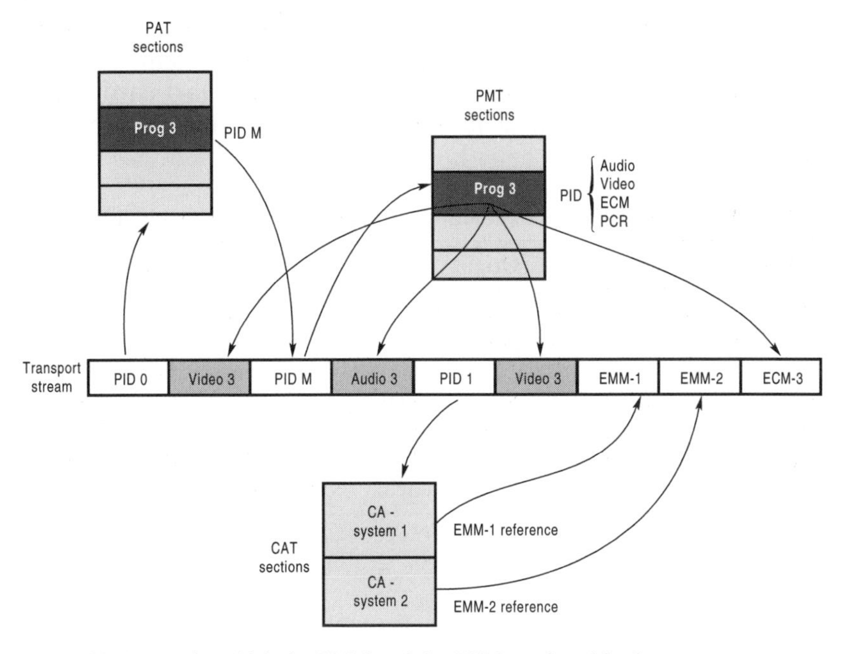

Fig. 5.30 Process by which the EMM and the ECM are found in the transport system

packets carrying the Programme Map Table sections (PMT) which indicate—in addition to the PID of the packets carrying the video and audio PES and the PCR—the PID of packets carrying the ECM.

The Conditional Access table (CAT), rebuilt from sections in packets with PID = 1, indicates which packets carry the EMM for one (or more) access control system(s).

From these pieces of information, and the user_key contained in the smart card, the descrambling system can calculate the control_word required to descramble the next series of packets (PES or transport depending on the scrambling mode).

The above-described process is indeed very schematic and the support containing the user_key and the real implementation of the system can vary from one operator to another. It is, of course, obvious that the details of these systems are not in the public domain.

Main conditional access systems

Table 5.1 indicates the main conditional access systems and associated scrambling algorithms used by the most important European pay-TV providers. Most of these systems use the DVB-CSA scrambling algorithm with a descrambler and conditional access software integrated in the receiver, the access rights being contained in the subscriber's smart card. Systems allowing pay-per-view are often equipped with a second smart card reader for a banking card and a modem for programme ordering and payment.

Table 5.1 The main European conditional access sytems for digital TV

System	Scrambling	Origin	Service provider(s)
CryptoWorks	DVB-CSA	PHILIPS	Viacom etc.
IrDETO	DVB-CSA	Nethold	DF1, Telepiu, Multichoice etc.
Mediaguard	DVB-CSA	SECA (Canal +)	Canal+, Canal Satellite
Nagravision	DVB-CSA	Kudelski S.A.	EchoStar, Via Digital etc.
Viaccess	DVB-CSA	France Telecom	TPS, AB-Sat, SSR etc.
Videoguard	ICAM	News Datacom (NDS)	BSkyB

5.4 Error correction and RF modulation (DVB-S satellite standard)

Once the source coding operations have been performed (including multiplexing and, eventually, scrambling), a transport stream, made of 188 byte packets, is available for transmission to the end users via a radiofrequency link (satellite, cable, terrestrial network). These transmission channels are, unfortunately, not error free due to many disturbances (noise, interference, echoes, and so on) that can combine with the useful signal. However, a digital TV signal, in which almost all redundancy has been removed, requires a very low Bit Error Rate (**BER**) for good performance (BER of the order of 10^{-10} to 10^{-12}). It is therefore necessary to take preventive measures before modulation in order to allow detection and, as far as possible, correction in the receiver of most errors introduced by the physical transmission channel. These measures, the majority of which consist of reintroducing a *calculated* redundancy into the signal, are called Forward Error Correction (**FEC**), or channel coding.

Figure 5.31 is a schematic representation of the channel coding process.

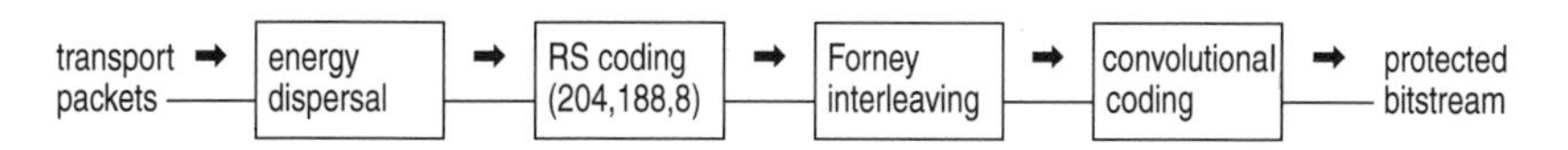

Fig. 5.31 The main steps of Forward Error Correction at the transmitter

5.4.1 Energy dispersal

The purposes of energy dispersal are to avoid a long series of '0's or '1's and to make the signal quasi-random to ensure energy dispersal of the radiofrequency spectrum after modulation (uniform energy distribution in the transmission channel). This is achieved by scrambling the data using a pseudo-random sequence. In order to locate the beginning of the sequence, the synchronization byte of the first packet of the group of eight packets to which the sequence is applied is inverted (47h becomes B8h). Apart from this, the scrambling does not affect the synchronization bytes themselves; the scrambling sequence must be applied even without an input signal in order to avoid transmission of a pure carrier.

5.4.2 Reed–Solomon coding ('external' coding)

In order to be able to correct most errors introduced by the physical transmission channel, we have already indicated that it is necessary to introduce some form of redundancy, thus allowing the detection and (within limits) correction of transmission errors in order to obtain a Quasi-Error Free (QEF) channel.

The first error correction coding layer, called 'outer coding', is used with all DVB specified transmission media; a second layer called 'inner coding' is used only in satellite and terrestrial transmissions. The outer coding is a *Reed–Solomon code* RS(204,188, T = 8) which, in combination with the Forney convolutional interleaving that follows it, allows the correction of burst errors introduced by the transmission channel. It is applied individually to all packets, including the synchronization bytes.

The RS(204,188, T = 8) coding adds 16 *parity bytes* after the information bytes of the transport packets, which therefore become 204 bytes long. In this way, up to 8 erroneous bytes per packet can be corrected. However, if there are more than 8 erroneous bytes in the packet, it will be indicated as erroneous and not correctible, and it is up to the rest of the circuitry to decide what to do with it.

Figure 5.32 shows the format of the protected transport packets.

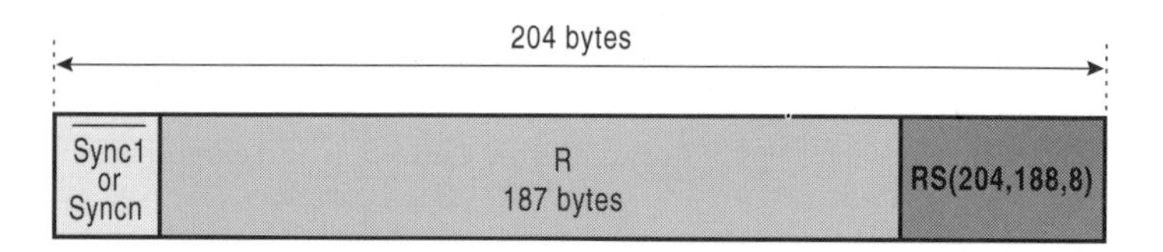

Fig. 5.32 A transport packet after Reed–Solomon coding

5.4.3 Forney convolutional interleaving

The purpose of this step is to increase the efficiency of the Reed–Solomon coding by spreading over a longer time the burst errors introduced by the channel, which could otherwise exceed the correction capacity of the RS coding (8 bytes per packet). This process, known as Forney convolutional interleaving, is illustrated in Fig. 5.33.

A burst of errors, after temporal reordering in the receiver, will be spread over two successive packets, and, most of the time, will stay within the correction capacity of the Reed–Solomon coding.

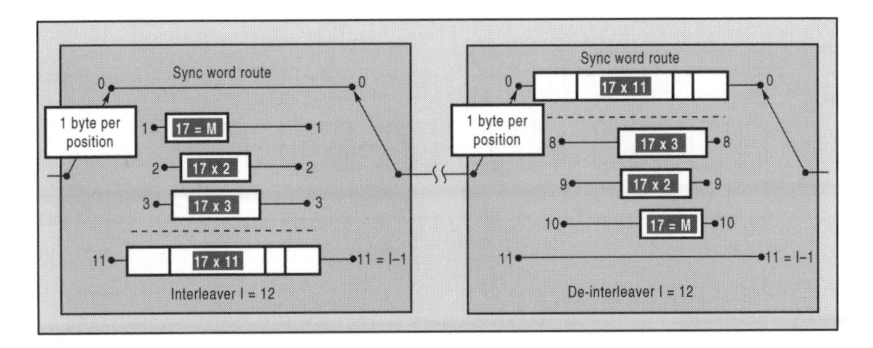

Fig. 5.33 Forney convolutional interleaving/de-interleaving (source: prETS 300 421)

The synchronization byte (inverted or not) always follows the path of index $j = 0$ in order to allow tracking.

5.4.4 Convolutional coding (inner coding)

For satellite transmissions, channel coding includes an additional operation known as 'internal coding' (sometimes called Viterbi coding, named after the algorithm used for decoding it). Internal coding aims to correct as many errors as possible that are due to the low signal-to-noise ratio.

The error correction allowed by this convolutional type coding complements the correction brought by interleaving plus Reed–Solomon coding. Its purpose is to obtain, starting from a Bit Error Rate (BER) of the order of 10^{-2} at the output of the QPSK demodulator, a BER of 2×10^{-4}, after Viterbi decoding, which allows a 'quasi error free' bitstream after Reed–Solomon decoding (a BER of the order of 10^{-10} to 10^{-11} after RS decoding).

Figure 5.34 illustrates this coding process (which creates two binary streams from the original stream) with the parameters of the DVB standard.

The strong redundancy (100%) introduced by the basic convolutional coding (code rate $R_c = 1/2$) allows very powerful error correction. This can sometimes be necessary with a very low signal-to-noise ratio (SNR) at the input of the receiver, but it reduces by a factor of two the spectral efficiency of the channel. In this case, the X and Y output streams from the convolutional encoder are directly applied (after filtering) to the I and Q inputs of the QPSK modulator for a satellite transmission, and the useful bit-rate of the channel is half the transmitted bit-rate. However, convolutional coding allows this redundancy to be

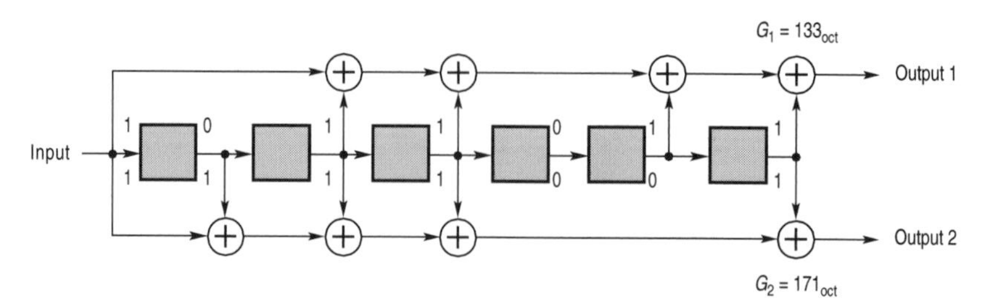

Fig. 5.34 Principle of the DVB-S convolutional coder

lowered by means of *puncturing* the outputs of the convolutional encoder. This consists of taking, at predefined times, only one of the two simultaneous bits to obtain one of the code rates specified by the DVB standard (R_c = 2/3, 3/4, 5/6 or 7/8, the 'FEC' published in satellite magazines). These numbers represent the ratio between the useful (input) and transmitted (output) bit-rates. Puncturing increases the capacity of the transmission channel at the expense of a reduction of the correction efficiency of the convolutional code.

The choice of the code rate by the broadcaster will therefore be a trade-off between useful bit-rate and reception antenna size for given transponder power and service area.

5.4.5 Modulation of DVB satellite signals

Once the channel coding operations have been carried out, we have a data stream ready to be used for modulation of a carrier for transmission to the end users.

For satellite transmissions in the Ku band, the channel width is most often between 26 MHz (Astra 1A–1D and 1G) and 36 MHz (Telecom), 33 MHz being the most common value for direct-to-home services (ASTRA 1E and 1F, Eutelsat 'Hot Birds'). Digital transmissions have inherited this situation since both types of transmissions will have to co-exist on the same satellite and, in addition, the new digital signals have to be compatible with most existing distribution equipment.

QPSK modulation (DVB-S)

The simplest digital modulation schemes use a direct modulation of the carrier by the bitstream, representing the information to be transmitted, either in amplitude (**ASK**, Amplitude Shift Keying) or in frequency (**FSK**, Frequency Shift Keying). In order to increase the spectral efficiency of the modulation process, different kinds of Quadrature Amplitude Modulations (QAMs) can be used. Figure 5.35 schematically shows the process of quadrature modulation and demodulation.

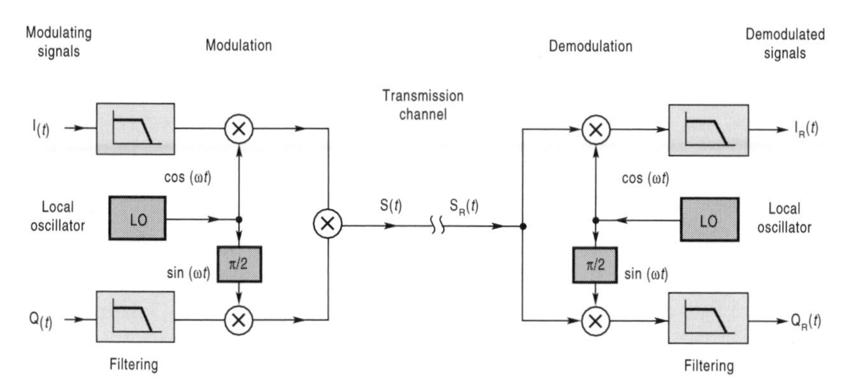

Fig. 5.35 The basic quadrature modulation/demodulation process

The I signal modulates an output of the local oscillator and the Q signal modulates another output in quadrature with the first (out of phase by π/2). The result of this modulation process can be represented as a *constellation* of points in the I, Q space. This constellation represents the number of *states* that depend on the values that I and Q can take. Figure 5.36 shows the ideal constellation of the QPSK modulation (four states).

The information carried by the bits, which define the amplitude of the modulating I and Q signals, is called the *symbol* (coded on n bits if I and Q are each coded on $n/2$ bits). The constellation thus has 2^n points.

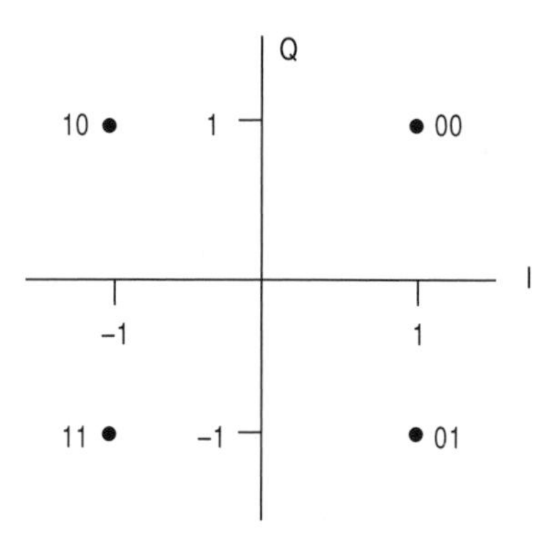

Fig 5.36 Constellation of a QPSK signal

To increase spectral efficiency, it is obvious that the number of bits per symbol has to be as high as possible, since the symbol rate (which determines the bandwidth) is equal to the bit-rate (in symbol/second divided by the number of bits per symbol, 2 in the case of QPSK). However, the noise amplitude relative to the I and Q signals received (reflected in the signal-to-noise ratio S/N or C/N) limits the possibility of distinguishing between the various possible states of I and Q: if the 'clouds' of points surrounding each theoretical point of the constellation overlap, it will not be possible to distinguish between the points. Thus, the greater number of different states the modulation scheme uses, the higher the signal-to-noise ratio will have to be.

Taking into account the low C/N values obtained with satellite reception, QPSK modulation (two bits per symbol, corresponding to four states) has been found to be the practical maximum. In this case, I (In phase) and Q (Quadrature) signals are the output of the convolutive coder (after the puncturing stage if $R_c \neq 1/2$).

Nyquist filtering

Without filtering, the frequency spectrum of digital signals is theoretically infinite, which would imply an infinite bandwidth for their transmission; this is, of course, not possible. Therefore, appropriate filtering will be required to limit the required bandwidth. The most commonly used filter is called a raised cosine filter, or more simply, a *Nyquist filter*.

In order to optimize the bandwidth occupation and the signal-to-noise ratio, filtering is shared equally between the transmitter and the receiver, each of which comprises a half-Nyquist filter.

This filter is characterized by its *roll-off factor* α, which defines its steepness (Fig. 5.37). For a signal with a symbol period T (symbol frequency or symbol rate $1/T$) the bandwidth B occupied after Nyquist filtering with a roll-off α is given by the relation: $B = (1 + \alpha)/2T$. Figure 5.37 shows the response curve of the Nyquist filtering (normalized to the symbol rate $1/T$) for three values of the roll-off factor (0.2, 0.35, 0.5).

The DVB-S standard prescribes a value of 0.35 for the roll-off factor. Consequenty, the RF bandwidth after QPSK modulation (theoretical channel width) will be:

$$B = (1 + \alpha) \times 1/T$$

For a 33 MHz channel, the theoretical maximum symbol rate will thus be: 33/1.35 = 24.4 Msymb/s.

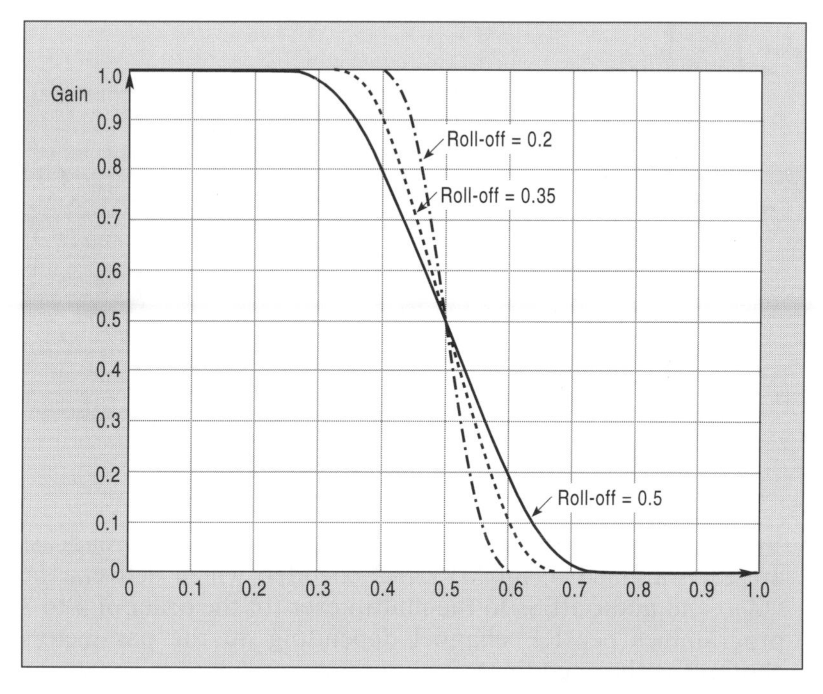

Fig. 5.37 Frequency response of the Nyquist filter for three values of the foll-off factor

In practice, it is possible—without a major drawback (other than an apparent C/N degradation of 0.2 dB)—to use a ratio of only 1.20 between the channel width and symbol rate, which leads to the most common symbol rates of 27.5 Msymb/s, which is used with 33 MHz channels (e.g. Astra 1E and 1F and Eutelsat's 'Hot Birds'), and 22 Msymb/s for 26 MHz channels (Astra 1G).

5.5 Reception of digital TV

5.5.1 Global view of the transmission/reception process

As an introduction to the description of an Integrated Receiver/ Decoder (IRD or, more commonly, set-top box), we describe the various processing steps that a TV signal has to follow from its source to its display on the end-user's screen.

The upper part of Fig. 5.38 illustrates the steps on the transmission side, which has to deliver a multiplex of MPEG-2 programmes on the one RF channel:

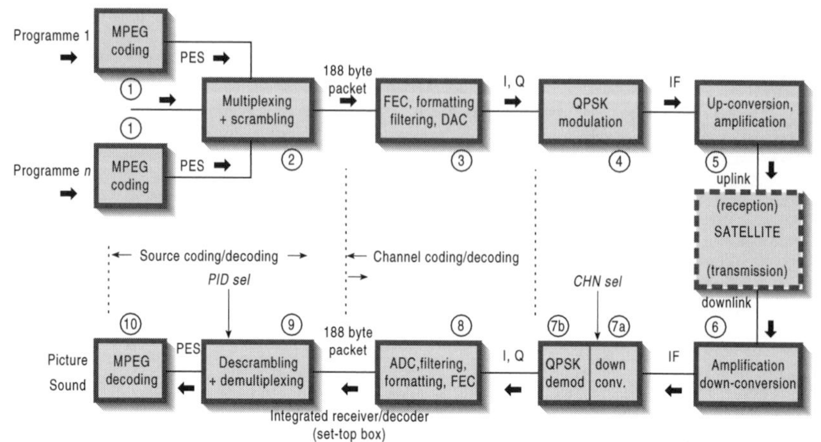

Fig. 5.38 The complete DVB transmission/reception chain

1. Video and audio signals of the programmes to be broadcast are each applied to an MPEG-2 encoder, which delivers the video and audio PESs to the multiplexer (of the order of 4 to 8 programmes per RF channel depending on the parameters chosen for the encoding).
2. These PESs are used by the multiplexer to form 188 byte transport packets, which are eventually scrambled (CAT tables carrying conditional access information ECM/EMM are inserted in this case) as well as the PAT, PMT, PSI and DVB-SI tables for the electronic programme guide (EPG).
3. The RS error correction increases the packet length to 204 bytes; in addition, the convolutional coding multiplies the bit-rate by a factor of 1.14 (R_c = 7/8) to 2 (R_c = 1/2); formatting of the data (*symbol mapping*), followed by filtering and D/A conversion, produces the I and Q analogue signals.
4. The I and Q signals modulate in QPSK an IF (intermediate frequency) carrier of the order of 70 MHz.
5. This IF is up-converted into the appropriate frequency for its transmission to the satellite *transponder*.

This frequency change brings the frequency to the required value (of the order of 14 GHz) for the uplink to the satellite transponder where it will again be frequency converted for its diffusion (downlink) to the end users in the Ku band (from 10.70 to 12.75 GHz).

On the receiver side, the lower part of Fig. 5.38 shows the

complementary steps, which indeed happen in the reverse order of the transmission side.

6. Antenna: amplification and a first down-conversion take place in the Low Noise Converter (LNC), which changes the frequency to the 950 to 2150 MHz range (SAT-IF), and it is then led by a coaxial cable to the input of the IRD.
7. In the IRD, a second down conversion (7a) used for RF channel selection delivers an intermediate frequency of 480 MHz; the demodulation (7b) of this IF delivers the I and Q analogue signals.
8. After A/D conversion, filtering and reformatting of I and Q (symbol demapping), the Forward Error Correction recovers the transport packets of 188 bytes.
9. The demultiplexer selects the PES corresponding to the programme chosen by the user, which is eventually descrambled with the help of the ECM, EMM and user key (smart card).
10. The MPEG-2 decoder reconstructs the video and audio of the desired programme.

5.5.2 Composition of a digital TV Integrated Receiver/ Decoder (IRD)

We will now go into a little more detail, without, however, going down to the level of electrical diagrams which, due to their very high complexity, would offer little of interest.

The technological evolution in this field is extremely rapid, because it has to take advantage, as quickly as possible, of the continuous progress of integration. Thus, the lifetime of a generation of hardware will probably not exceed one year. In 1998, however, the cost of a digital receiver was still substantially higher than the cost of its analogue counterpart.

The block-diagram of Fig. 5.39 represents the main functional blocks of a satellite IRD. It does not, however, necessarily correspond to the partitioning used by all chip makers for realization of their ICs. This partitioning can vary substantially from supplier to supplier, and depends on the integration level, which increases quickly between two succcessive generations of hardware.

The signals received from the satellite (frequencies ranging from 10.7 to 12.75 GHz, are amplified and down-converted (in two bands) into the 950–2150 MHz range by the Low Noise

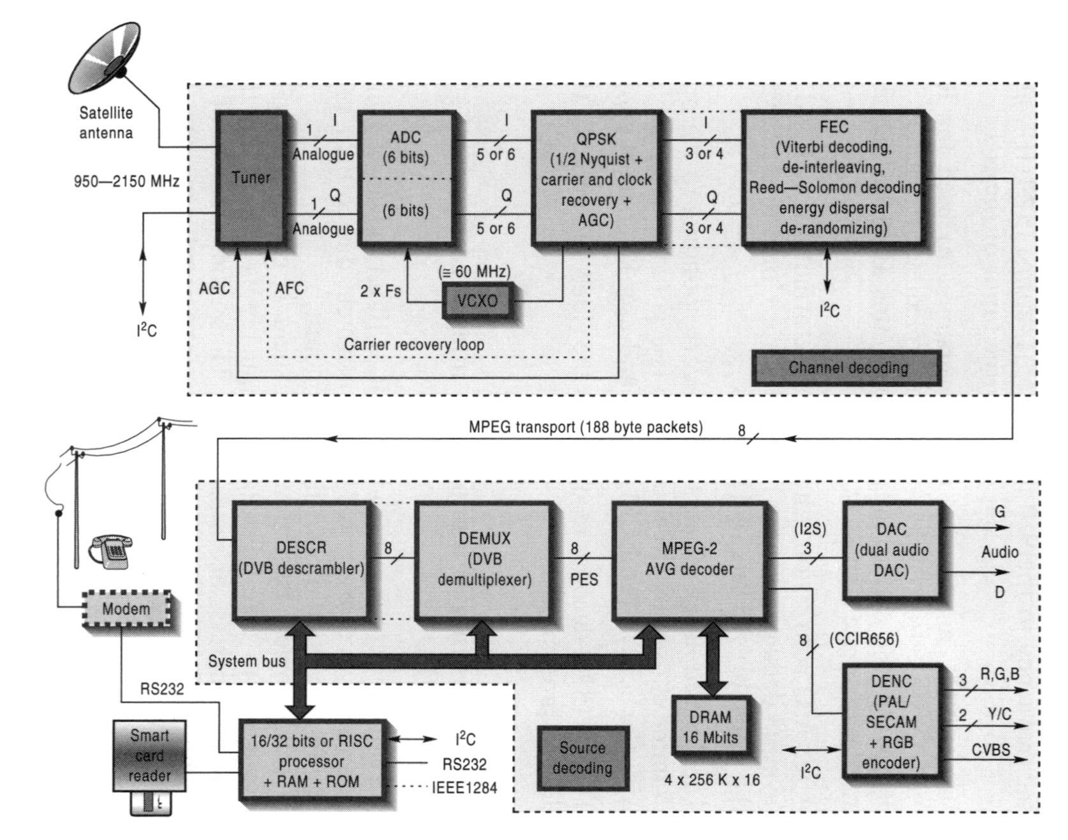

Fig. 5.39 Block diagram of a DVB satellite receiver (1995/96 generation) (the dotted lines represent functions that are sometimes combined)

Converter (LNC or LNB) located at the antenna focus, and applied to the IRD's input.

The *tuner*, generally controlled by an I^2C bus, selects the required RF channel in the 950–2150 MHz range, converts it into a 480 MHz IF and realizes the required selectivity by means of a *Surface Acoustic Wave filter* (SAW); the signal is amplified and coherently demodulated according to the 0 and 90° axis to obtain the analogue I and Q signals. Recovery of the carrier phase required for demodulation is carried out in co-operation with the next stages of the receiver, which lock the phase and the frequency of the local oscillator by means of a Carrier Recovery Loop.

The ADC (Analogue-to-Digital Converter) receives the analogue I and Q signals, which it then converts at twice the symbol frequency F_{SYMB} (in Europe most often 27.5 Msymb/s or 22 Msymb/s). In most cases, this is performed by means of a dual ADC with 6 bit resolution, which has to work with a sampling frequency of more than 60 MHz. Here again, the sampling frequency is locked to the symbol frequency by means of a phase locked loop (clock recovery loop).

The QPSK block, in addition to its functions in carrier and clock recovery loops mentioned before, carries out the half-Nyquist filtering, which is complementary to the one applied on the transmitter side, to the I and Q signals. Digitized I and Q signals are delivered to the next functional block (FEC).

The FEC block (Forward Error Correction) distinguishes, by means of a majority logic, the '0's from the '1's and realizes the complete error correction in the following order. Viterbi decoding of the convolutional code, de-interleaving, Reed–Solomon decoding and energy dispersal de-randomizing; the output data are the 188 byte transport packets that are generally delivered in parallel form (8 bit data, clock and control signals, of which one generally indicates uncorrectable errors).

The DESCR block (descrambler) receives the transport packets and communicates with the main processor by a parallel bus to allow quick data transfers. It realizes the selection and descrambling of the packets of the required programme under the control of the conditional access device. This function is sometimes combined with the demultiplexer.

The DEMUX (demultiplexer) selects, by means of programmable 'filters', the PES packets corresponding to the programme chosen by the user.

The audio and video PES outputs of the demultiplexer are applied to the input of the MPEG block, which generally combines MPEG audio and video functions and the graphics controller functions required, among other things, for the electronic programme guide (EPG). MPEG-2 decoding requires generally at least 16 Mbits of **DRAM** or **SDRAM**.

Video signals reconstructed by the MPEG-2 decoder (digital YUV signals in CCIR 656 format) are then applied to a digital video encoder (DENC), which ensures their conversion into analogue RGB + sync. for the best possible display quality on a TV set via the SCART/PERITEL connector, and PAL or NTSC or SECAM (composite and/or Y/C) mainly for VCR recording purposes.

Decompressed digital audio signals in **I²S** format or similar are fed to a dual digital-to-analogue converter (DAC) with a resolution of 16 bits or more which delivers the analogue left and right signals. A digital (optical or electrical) audio output is sometimes provided.

The whole system is controlled by a powerful 32 bit microprocessor (often a RISC processor) which controls all the circuitry, interprets user commands from the remote control, manages the smart card reader(s) and the communication interfaces generally available. The software amounts to many hundreds of kilobytes, which are located in a **Flash-EPROM** in order to permit eventual updates during the lifetime of the product (off air or via the communication ports).

The conditional access device generally includes one or two smart card readers (one might be for a banking card, for example). In the case of a detachable conditional access module using the DVB-CI common interface (signified by the presence of PCMCIA slots), the conditional access circuits and software, and in some cases a descrambler and a smart card reader, are located in the detachable PCMCIA module.

The IRD can communicate with the external world (PC, modem, etc.) by means of one or more communication ports. From the simplest (serial RS232) to the fastest (parallel **IEEE1284**) these ports—as well as a telephone line interface (via an integrated modem)—are the necessary connection points required for interactivity and access to special services (pay-per-view, teleshopping, access to networks and so on).

The digital receiver is often equipped with a *loop-through* functionality in its tuner which allows the connection of a second receiver (generally analogue) to the same dish. This ensures the signal path, and leaves the LNB control and supply to the analogue receiver when the IRD is in standby condition. In this case, the connection of the IRD to the TV is generally carried out

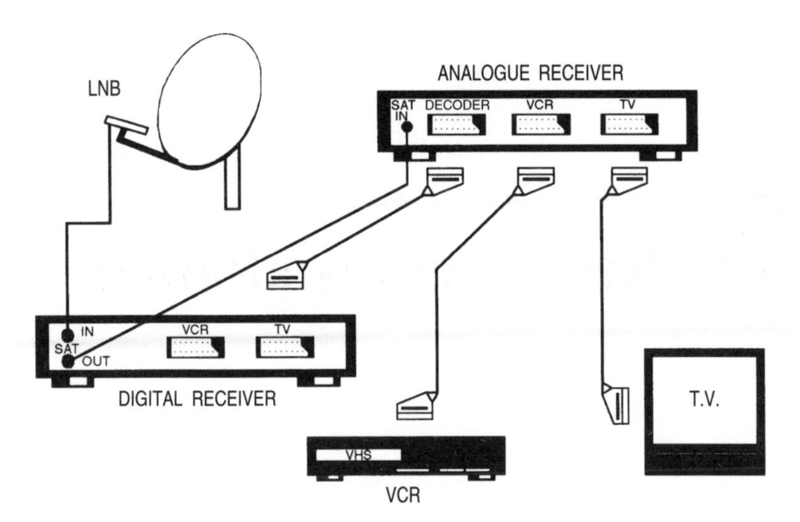

Fig. 5.40 Example of interconnection of an IRD, a VCR and a TV through the analogue SAT receiver

through the 'decoder' SCART connector of the analogue receiver, which can play the role of a mini AV switchboard between the TV, the VCR and the IRD if it is equipped with three SCART connectors (Fig. 5.40).

6 Receiving installation

6.1 The antenna

As we have already seen, these days the antenna is most often the parabolic offset focus type, and it can either be fixed or motorized. Its dimensions are determined by the minimum required C/N ratio, itself determined by the percentage of the time during which reception should be possible, the EIRP of the satellite(s) to be received, the efficiency of the antenna and the LNB characteristics.

We saw in Chapter 3 the relation between C/N, G/T and EIRP:

$$\text{C/N [dB]} = \text{EIRP} - A_i + G_R - 10 \log (kT_S\, B) = G/T + [\text{EIRP} - (A_i + 10 \log kB)] \qquad (3.1)$$

from which we get:

$$G/T_{min} = \text{C/N}_{min} - [\text{EIRP}_{min} - (A_{i\,max} + 10 \log kB)]$$

It is consequently possible to find the minimum G/T of the installation by fixing the following parameters:

- the minimum EIRP of the satellite(s) to be received,
- the worst case atmospheric losses which determine $A_{i\,max}$,
- the bandwidth B of the receiver which determines 10 log kB.

It will then be possible to deduce the antenna dimensions by knowing:

- the efficiency of the antenna which determines *G*,
- the noise figure of the LNB which determines *T*.

Table 6.1 gives the dimensions of an offset antenna (assuming an efficiency of 60%) as a function of the EIRP of the satellite with an LNB noise figure of 1.3 dB in order to guarantee a minimum C/N of 11 dB for 99.75% of the time, in a 27 MHz channel in the Ku band (see the calculation details in Appendix A.3).

Table 6.1 Figure of merit and antenna dimensions as a function of the EIRP (analogue TV)

EIRP (dBW)	40	42	44	46	48	50	52	54	56
G/T min	24.2	22.2	20.2	18.2	16.2	14.2	12.2	10.2	8.2
Φ (cm) min	250	200	160	125	100	80	63	50	40

If the antenna is equipped with more than one LNB (multi-satellite reception), the antenna efficiency will be reduced and the antenna dimension will have to be increased to compensate for this additional loss.

6.2 The antenna–receiver link

Similar to a conventional (terrestrial) TV antenna, the LNB is connected to the satellite receiver by means of a coaxial cable of 75 Ω impedance; its length can be up to some tens of metres, depending on the constraints of the place of installation. There are, however, two important differences with conventional TV, which result in additional constraints:

- the output frequencies of the LNB (Satellite Intermediate Band, from 950 to 2150 MHz) are much higher than those of terrestrial TV (VHF/UHF from 47 to 860 MHz),
- the coaxial cable also carries the supply of the LNB (13/18 V, approximately 200 mA), the bandswitching control signal (22 kHz), and sometimes other switching signals (60 Hz, DiSEqC, etc.).

6.2.1 Coaxial cable

As its name implies, coaxial cable is made of a central copper conducting wire (core) surrounded by a dielectric cylindrical tube, itself surrounded by shielding copper foil (although this is sometimes absent) and a copper braid, all being protected by an insulating sleeve. The ratio between the central conductor core diameter and the internal diameter of the shield, and the dielectric characteristics, determine the characteristic impedance of the cable.

This impedance is standardized to 75 Ω for all TV applications (terrestrial as well as satellite); it has to be respected to ensure correct matching in order to avoid reflections and additional losses between the LNB and the receiver input. Owing to the higher frequencies used, losses due to the cable are more important than in terrestrial TV, especially in the upper part of the SAT-IF band above 2000 MHz. In particular, in the case of a long cable (more than 20 m), careful attention should be paid to the quality of the cable, of which the loss is specified for a length of 100 m at different frequencies (often 1000 and 2000 MHz).

Depending on the quality of the cable, its attenuation will be of the order of 20 to 30 dB for 100 m at 1000 MHz, and will increase by 10 to 15 dB at 2000 MHz. In addition, since the cable carries the supply current of the LNB (of the order of 200 mA), the voltage drop between the receiver and the LNB has to be less than 1 V in order to avoid any problems of polarization switching: this depends mainly on the diameter of the core conductor, which should not be too small, especially if the cable is long. The cable will also have to be compatible with the type of connectors used (so-called F connectors).

6.2.2 Connectors

Until the beginning of the 1990s, some European receivers used the traditional 9.5 mm IEC connector used in TV, but the 'F' connector is now the de-facto standard, for the LNB output as well as for the antenna input of the receiver and any other optional devices (amplifier, switches, etc.) inserted between the LNB and the receiver.

The connector used for satellite antennas is known as 'type F', and this type of connector uses the central wire (core) of the coaxial cable instead of a true contact pin (except for special models intended for use with very large or very small cables). As a result of this, the cable will have a rigid monostrand core, with a diameter not less than 0.7 mm (to ensure a good contact) and not more than 1 mm (in order not to deteriorate the female connector). Many types of F connectors exist, for crimping or screwing on the cable, and can be adapted to the dimensions of the coaxial cable.

The model for screwing onto the cable will be preferred by the occasional user, since it ensures a quick and robust connection without requiring any special tool, and is adapted to the most common cables having an external diameter between 6 and 7 mm.

For the professional working on important installations, crimp models (which exist for cables from 5 to 15 mm in diameter) are preferrable, since they are quicker to mount, but they require a special tool for correct mounting.

For large cables with a core that can be 2 mm in diameter, there are special connectors that are equipped with a central contact of approximately 1 mm diameter.

For connectors used outdoors (e.g. on the LNB), some precautions are necessary to ensure waterproofness (rubber cap, silicon paste or special adhesive tape) in order to avoid any oxidation which could result in false contacts, or even cutting of the core conductor after months or years of exposure to the elements.

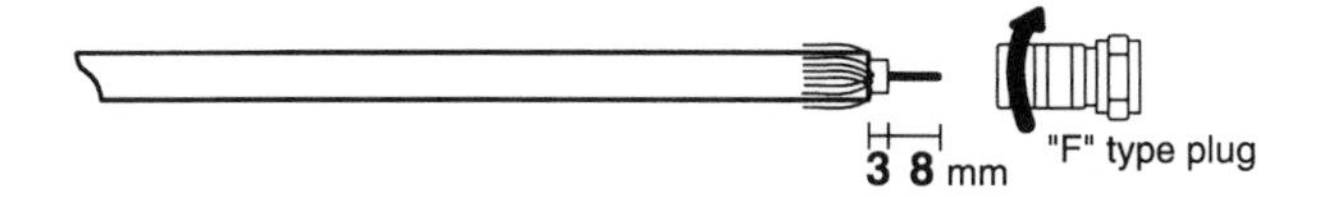

Fig. 6.1 Mounting of an F connector (screwed model)

6.3 Antenna switching devices

The need to switch many antennas (or many LNBs placed on the same antenna) was present from the beginning of individual satellite reception, and numerous non-standard devices have been, and (for some of them) still are, in use.

One of the simplest switching methods, when only two antennas are to be switched, is to have two antenna inputs on the receiver, which is the case for many analogue receivers from the mid-range and higher; in this case, antenna switching is carried out in the receiver, but this requires two cables from the antennas, which is often impractical if not impossible.

Another method, not much more practical, is to have a switch near the antennas, remotely controlled by means of a logic (DC) level (the most common being a 0/12 V control signal); it requires only one coaxial cable, but at least one wire carrying the control signal (the return can be ensured by the coaxial cable shielding). In order to avoid the need for a special wire to carry this information, a means of using coaxial cable has been sought, the most obvious method being to superimpose a 'tone' on the LNB supply voltage; the 22 kHz tone was the first de-facto standard for this purpose and was used at the beginning of the 1990s. This tone was, however, used for bandswitching when the universal LNB appeared; so a second tone of lower frequency (60, 175 or 400 Hz), easily separable from the 22 kHz tone, was added. The 60 Hz value, if not really a standard, is the most common frequency.

These systems, apart from the fact that they are not standardized and often require external 'tone injectors', are limited to two antennas, which is not always sufficient. This is why Eutelsat proposed in 1995 a new switching system named **DiSEqC** (Digital Satellite Equipment Control): it is a real 'control bus' using the coaxial cable as a vehicle, and its principle consists of modulating the 22 kHz tone by the messages to be carried to the antenna system. It has been designed to be compatible with the switching principles of the 'universal' LNB (13/18 V for the polarization and 22 kHz for the band), and its use is royalty free.

6.3.1 The DiSEqC bus and its different levels

The DiSEqC specification foresees many levels (*ToneBurst* or *mini-DiSEqC*, DiSEqC 1.0, DiSEqC 1.1, DiSEqC 1.2, DiSEqC 2.0 and so on), which are responses to the increasing requirements and allow a progressive rollout.

DiSEqC has three implementation levels: 'ToneBurst', DiSEqC 1.*x*, DiSEqC 2.*x* and above.

- The 'ToneBurst' (or mini-DiSEqC) is a simple extension to the 22 kHz tone intended to be implemented in the short-term in receivers or external accessories to allow dual band reception of a maximum of two satellites. The implementation of ToneBurst only in equipment does not give the right to display the DiSEqC logo (see Fig. 6.16, page 151) but only to the 'DiSEqC compatible' appellation.
- The first DiSEqC levels (DiSEqC 1.0 to 1.2) use a unidirectional communication between the receiver and the DiSEqC peripherals, and allow a simple implementation in the receiver (which is mainly software based) as well as in the peripherals. DiSEqC 1.1 adds new commands compared with DiSEqC 1.0 and foresees the repetition of commands. DiSEqC 1.2 allows the control of an antenna positioner.

- Higher levels (DiSEqC 2.0 and above) are more ambitious and aim, in the long term, at a more or less 'plug and play' installation through a dialogue between the receiver and the slave peripherals by means of a kind of 'return channel' from the peripheral to the receiver.

For each DiSEqC level, a minimal functionality has been defined by Eutelsat for receivers and peripherals (Table 6.2):

Table 6.2 Minimum functionality of receiver for the various DiSEqC levels

Level	Communication	Minimum receiver functionality
ToneBurst	unidirectional	14/18 V + 0/22 kHz + control of a switch (Sat A/Sat B)
DiSEqC 1.0	unidirectional	Idem ToneBurst + Control of 4 'committed' switches
DiSEqC 1.1	unidirectional	Idem 1.0 + additional controls + repetition (× 3)
DiSEqC 1.2	unidirectional	Idem 1.1 + positioner control + polarizer control (optional)
DiSEqC 2.x	bidirectional	Idem 1.x + reading/processing of return data from peripherals

6.3.2 DiSEqC operation

DiSEqC is a bus system with a unique *master* (the satellite receiver), which is the only unit that can take the initiative of communication on the bus, and one or more *slave* peripherals (often in the neighbourhood of the antenna), which can only react and respond to a command. DiSEqC messages are obtained by PWK modulating (Pulse Width Keying, see Fig. 6.2) the duty cycle of the 22 kHz which can take two values: '0' corresponds to 2/3 duty cycle and '1' to 1/3 duty cycle. The bit period is 33 cycles of 22 kHz (1.5 ms ± 20% taking into account the 22 kHz tolerance). Each DiSEqC message must be followed by at least 6 ms of 'silence'.

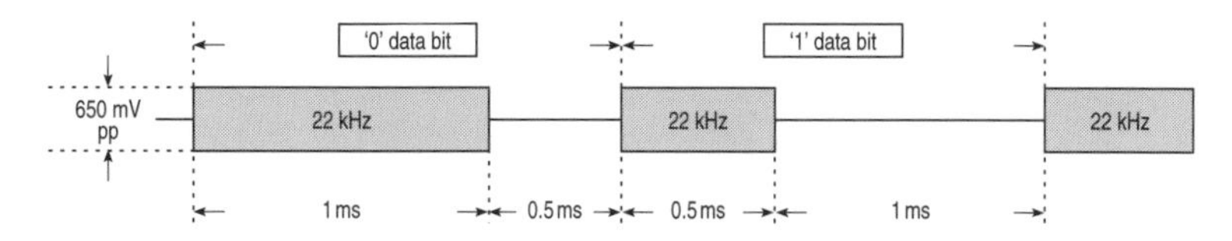

Fig. 6.2 Definition of the DiSEqC data bit

The 22 kHz level superimposed on the LNB supply voltage is nominally 650 mV_{pp}. The master should have a source impedance of 15 Ω at 22 kHz in order to be able to receive the response from the slaves (for levels 2.0 and above).

DiSEqC is specified in a such a way as to be compatible with existing hardware (see Fig.6.3):

- in order to be compatible with existing peripherals (14/18 V, 0/22 kHz, ToneBurst) new receivers must first apply the appropriate voltage for polarization control and the 22 kHz tone if required, then, after a short 'silence', the true DiSEqC message is followed by the ToneBurst;
- in order that new DiSEqC peripherals be at least partly controllable by existing (non-DiSEqC) receivers, they will initially have to respond to 'traditional' 14/18 V and 22 kHz controls.

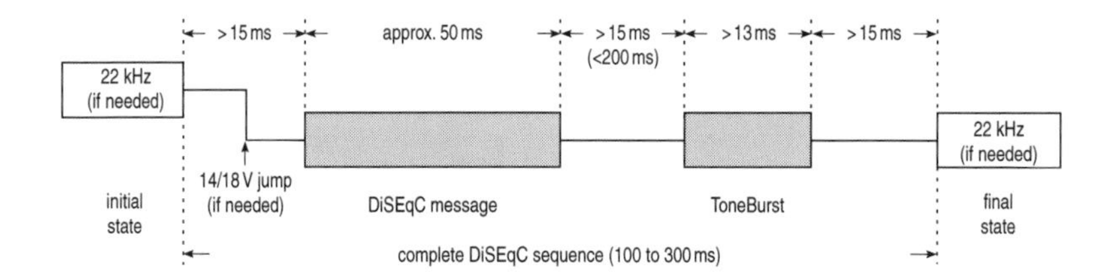

Fig. 6.3 DiSEqC sequence compatible with a Universal LNB and ToneBurst equipment

DiSEqC 'ToneBurst'

The ToneBurst consists of a either a plain 22 kHz burst of 12.5 ms (satellite A selection) or of a sequence of nine '1' s (1/3 duty cycle) for satellite B selection (see Fig. 6.4).

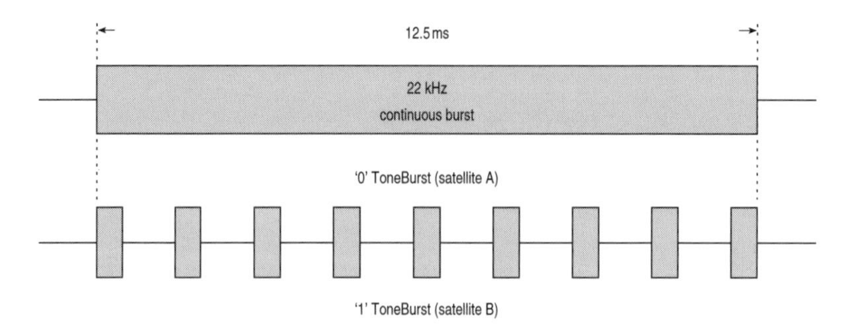

Fig. 6.4 Definition of the ToneBurst

DiSEqC 1.0 and higher

DiSEqC messages are made by one or more bytes (MSB first) each followed by an odd parity bit in order to be able to detect transmission errors.

- Commands from the *master* consist of a minimum of 3 bytes (framing, address, control), followed by a certain number of data bytes for some commands.

  The general format is illustrated below.

FRAMING	P	ADDRESS	P	CONTROL	P	DATA	P	DATA	P

  In order to keep the duration of a message within acceptable limits, a maximum of one byte of data is used with level 1.0, and a maximum of three for levels 2.0 and above.

- The response from a *slave* is at the master's discretion, and occurs only if they are both at level 2.0 or higher. It comprises a framing byte (used as an acknowledgement) optionally followed by one or more data bytes carrying the message content. Its format is shown below.

FRAMING	P	DATA	P	DATA	P

  The slave response must occur within 150 ms after the master's command ends. An error recovery method is not defined by the specification; the most usual strategy consists of repeating the message, but this is entirely at the master software's discretion.

6.3.3 Different fields of the DiSEqC message

Framing byte

This is the first byte of a master's command or a slave's response.

- The 4 MSBs are used for synchronization (*run-in*) and recognition (*framing*) of a DiSEqC message. They are currently fixed at '1110', but the last two bits might be affected later.
- The 5th bit is reserved (currently '0') as it might be used later.

- The 6th bit indicates the origin of the message: '0' if it comes from the master, '1' if it comes from a slave.
- The 7th bit is set to '1' if a response is expected, '0' otherwise. In the case of a slave response, it can be set to '1' to ask for a repeat of the message in the case of a transmission error.
- The 8th bit is set to '1' by the master if it is a retransmission. In the case of a slave response, it indicates the reason for non-execution of the command: '0' in the case of a parity error, asking for retransmission, '1' if the command is not supported by the slave.

The meanings of the framing bytes are described in Table 6.3; note that only the two first lines apply to level 1.0.

Table 6.3 Meaning of the framing byte

Hex.	Binary	Meaning
E0	1110 0000	master command, no response expected, 1st transmission
E1	1110 0001	master command, no response expected, 2nd transmission
E2	1110 0010	master command, response expected, 1st transmission
E3	1110 0011	master command, response expected, repetition
E4	1110 0100	slave response, repetition unnecessary, no error
E5	1110 0101	slave response, repetition unnecessary, command not supported
E6	1110 0110	slave response, repetition required, parity error
E7	1110 0111	slave response, repetition required, command not recognized

Address byte

The second byte sent by the master indicates the slave to which the message is destined; it is subdivided into two 'nibbles'. The MS (Most Significant) nibble indicates the family to which the slave belongs (LNB, switch, positioner etc.), the LS (Least Significant) nibble divides the family into sub-groups. For each nibble the value '0000' indicates a value for all families (MS nibble) or all sub-groups of a family (LS nibble). The address hex. 00 is recognized by all peripherals.

Table 6.4 indicates the addresses currently specified (addresses above hex. 20 are only used by levels 1.1 and above).

Control (or command) byte

This byte allows up to 256 basic commands, optionally complemented by one or more data bytes. DiSEqC specification (version

Table 6.4 Meaning of the address byte

Hex. address	Binary	Category
00	0000 0000	All families, all sub-groups (general call)
10	0001 0000	Switch (all types, including LNB or SMATV)
11	0001 0001	LNB
12	0001 0010	LNB with loop-through and switching
14	0001 0100	Switch without DC pass-through
15	0001 0101	Switch with DC and bus pass-through
18	0001 1000	SMATV
20	0010 0000	Polarizer (all types)
21	0010 0001	Linear polarizer control (full skew)
30	0011 0000	Positioner (all types)
31	0011 0001	Positioner (polar or azimuth type)
32	0011 0010	Positioner (elevation)
40	0100 0000	Installation support (all types)
41	0100 0001	Signal intensity adjustment support
6x	0110 xxxx	Reserved for address reallocation
70	0111 0000	Slave intelligent interface (all types)
71	0111 0001	Interface for subscriber controlled head station
Fx	1111 xxxx	Manufacturer extensions

4.2 of 25/02/1998) defines more than 50 commands, divided into three categories: mandatory (M), recommended (R) and suggested (S). For the 1.0 level, only three commands are taken into account, out of which one only is mandatory (38 'write N0'), the 'reset' command (00) being only mandatory when the receiver has a loop-through function (see Table 6.5).

Commands 38 and 39 define, by means of an additional byte, the status of the outputs of a slave microcontroller defined by

Table 6.5 The main commands of DiSEqC 1.0 to 1.2

Hex.	Status	From level	Bytes	Name	Function
00	M/R	1.0	3	Reset	Reset DiSEqC microcontroller (peripheral)
03	R	1.0	3	Power On	Supply peripheral
38	M	1.0	4	Write N0	Write port 0 (committed switches)
39	M	1.1	4	Write N1	Write port 1 (uncommitted switches)
48	R	1.2	4	Write A0	Polarizer control (skew)
58	M	1.1	5 or 6	Write Freq.	Tune to transponder frequency
60 to 6B	M	1.2	3 or 4	various	Positioner commands

Eutelsat and intended to be integrated in peripherals. Commands 20 to 2F (optional) control individually each of the microcontroller's outputs, but are of little interest since they are redundant compared with commands 38 and 39 followed by an appropriate data byte. From level 1.1 up, commands 38 (write N1) and 58 (write frequency) are mandatory. This final command, followed by 3 bytes indicating the transponder frequency (in BCD coding), is intended for use in **SMATV** systems using a user controlled frequency transposer in the head station. DiSEqC 1.1 also foresees the repetition of commands (up to two times after the first use) in order to allow control of cascaded switches. DiSEqC 1.2 adds the commands of a postioner and (optionally) of a polarizer.

Data bytes

Some commands must be followed by one or more data bytes. For instance, commands 38 and 39 must be followed by a byte, of which each bit at '1' of the MS nibble (clear) clears individually one of the four outputs of one of the two ports of the slave microcontroller; the LS 'nibble' (set) sets them individually in the same way.

CLEAR	SET
$b_7b_6b_5b_4$ = 1111 = F_h	$b_3b_2b_1b_0$ = xxxx = x_h

A DiSEqC 1.0 message intended to control an output port of the slave microcontroller will follow the general form below:

FRAMING (E0 or E1)	ADDRESS (10 to 18)	COMMAND (38 or 39)	DATA (Fx)

For command 38h, three of the four outputs of the 'committed' port are dedicated to a predefined function (the fourth one being freely usable), according to Table 6.6.

The functionality of command 39h depends on the use of the uncommitted switches made by the manufacturer.

With DiSEqC 1.1, command 58h allows us to send to a so-called *agile* transposer (an 'intelligent' peripheral located in the head-end of an SMATV system) its tuning frequency by means of three bytes (maximum), which represent the BCD coded frequency digits. The command format is as follows:

E0h	71h	58h	$\times 10^4$	$\times 10^3$	$\times 10^2$	$\times 10^1$	$\times 10^0$	$\times 10^{-1}$
Framing	Address	Command	Frequency (mHz)					

Table 6.6 Meaning of the bits of the data byte following command 38h

Bit	b_3		b_2		b_1		b_0	
Function	Option (1)		Satellite (2)		Polarization		Band	
State	1	0	1	0	1	0	1	0
Result	β	α	B	A	Horiz.	Vert.	High	Low

(1) example: α = satellites group ASTRA/Eutelsat, β = satellites group Telecom 2A/2B
(2) example: A = Astra, B = Eutelsat (group α), A = Telecom 2A, B = Telecom 2B (group β)

For instance, the frequency 11.850 GHz would be coded as follows:

1110	0000	0111	0001	0101	1000	0001	0001	1000	0101	0000	0000
E	0	7	1	5	8	1	1	8	5	0	0

The last byte containing only zeros can be omitted.

Figure 6.5 shows the sequence of a complete DiSEqC 1.1 message in the case of a programme change (SMATV mode with remote controlled transposer).

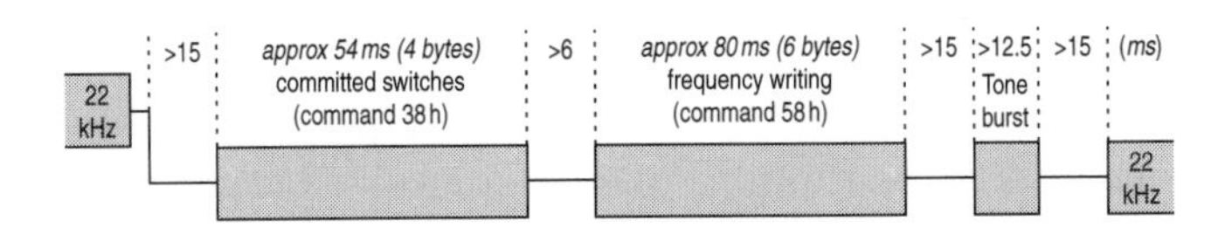

Fig. 6.5 Constitution of a DiSEqC 1.1 message with remote controlled tuning (SMATV)

6.3.4 Upper levels of DiSEqC

DiSEqC levels above 2.0 imply bidirectional communication between the master and the slave(s), which must always respond with at least a framing byte used as acknowledgement, followed—for certain commands—by one or more data byte(s).

One of the most interesting possibilities available to the receiver is to 'interrogate' the antenna configuration to which it is connected, at switching on, in order to configure itself. For instance, command 51h allows one to read the LNB characteristics (LO frequency among others). Other commands allow interrogation of the status of the committed and uncommitted switches, reading of analogue values such as the position of a motorized antenna, etc.

In the longer term, if DiSEqC is generalized to all peripherals (switches, LNBs, positioners, etc.), the retro-compatibility will not be required anymore; a 'true' bus structure will be possible (standardized address allocation, permanent bus and power supply distribution to all peripherals which will have a standby mode, and so on).

Note

DiSEqC specifications are available on the Eutelsat Internet site (http: //www.eutelsat.org).

6.4 Motorized antenna

A motorized antenna is the only practical solution for allowing reception of transmissions from a large number of different orbital positions on the geostationary arc. Many solutions are now available to the consumer.

6.4.1 'Classical' motorized antenna

The motorized antenna must ensure an automatic tracking of the geostationary arc when going from one satellite to another (automatic elevation correction with azimuth change). To achieve this goal, since the geostationary orbit is a circle centred on the polar axis, the antenna rotation will have to be centred on an axis parallel to the polar axis, and not around the vertical of its position (see Fig. 6.6), by means of a special unit called a *polar mount*.

The angle between the polar axis and the horizontal plane at the place of reception is equal to latitude λ of this location; its complement ($E = 90° - \lambda$) is the elevation of an object located at infinity in the equatorial plane, in the 'full south' direction at the place of observation. This is called the *polar axis elevation*.

Since the satellites are not located at infinity, but on a circle of approximately 42 000 km radius, the antenna elevation will have to be smaller than the elevation of the plane parallel to the equatorial plane; the angular difference for a satellite located 'full south' is called the *polar axis declination* δ (see Fig. 6.7).

However, since the rotational axis of the antenna is parallel to (but not identical to) the polar axis, there will not be perfect

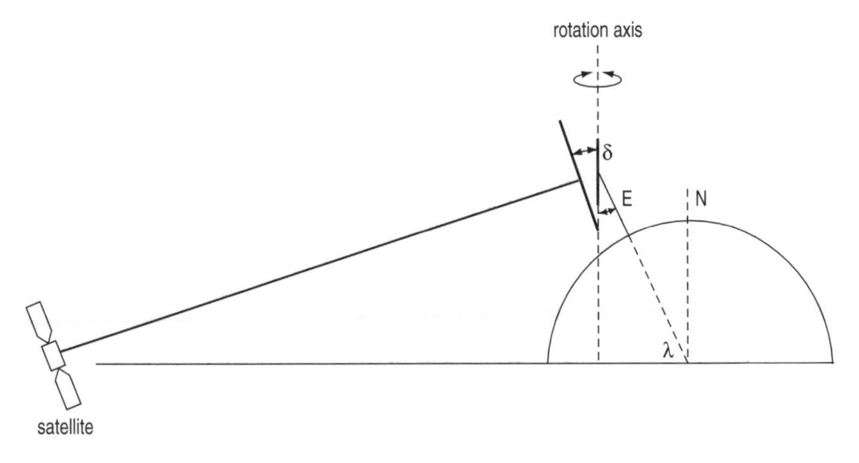

Fig. 6.6 Principle of the polar mount

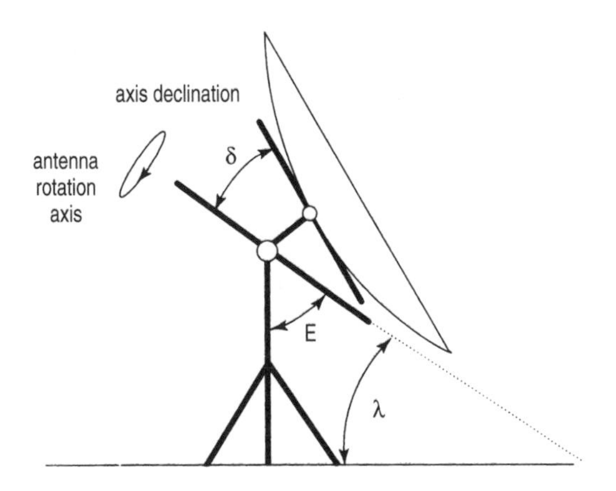

Fig. 6.7 Schematic representation of a polar mount antenna

tracking of the geostationary curve if one does nothing more than adjust the antenna with the polar axis declination of a satellite at 'full south' (Fig. 6.8): the circle tracked out will be smaller than the geostationary orbit, and will be centred on the projection of the place of reception onto the equatorial plane.

This can be corrected by slightly diminishing the polar axis elevation by a value ε (which lowers the antenna elevation), and by reducing by the same value the declination in order to compensate for a correct 'full south' adjustment. Between 40 and 55° latitude, this correction can be considered as constant ($\varepsilon \cong 0.65°$).

Table 6.7 gives the values of polar axis elevation and declination as a function of latitude (for values of latitude between 40 and 55°).

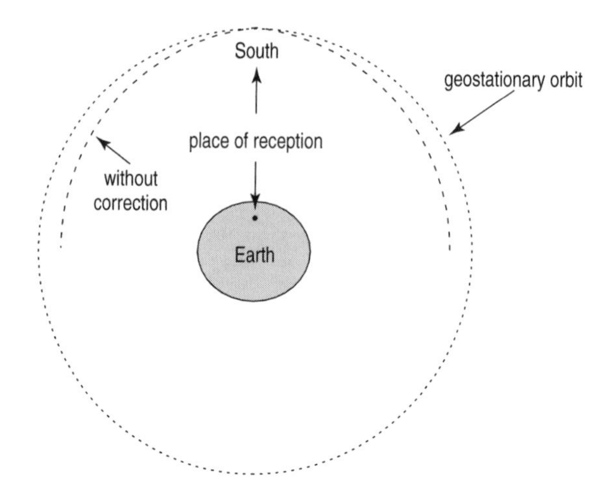

Fig. 6.8 Circle followed with a simple 'full south' adjustment

Table 6.7 Elevation and declination as a function of latitude (40 to 55°N)

Latitude (°)	Polar axis elevation (*E*)	SAT elevation at full south	Polar axis declination (δ)	Corrected polar elevation	Corrected polar declination
40	50	43.7	6.3	49.35	5.65
41	49	42.6	6.4	48.35	5.75
42	48	41.5	6.5	47.35	5.85
43	47	40.4	6.6	46.35	5.95
44	46	39.3	6.7	45.35	6.05
45	45	38.2	6.8	44.35	6.15
46	44	37.1	6.9	43.35	6.25
47	43	36.0	7.0	42.35	6.35
48	42	34.9	7.1	41.35	6.45
49	41	33.8	7.2	40.35	6.55
50	40	32.7	7.3	39.35	6.65
51	39	31.6	7.4	38.35	6.75
52	38	30.5	7.5	37.35	6.85
53	37	29.4	7.6	36.35	6.95
54	36	28.3	7.7	35.35	7.05
55	35	27.2	7.8	34.35	7.15

The rotation of a 'classical' motorized antenna is generally achieved by means of a jack, under the control of a positioner, integrated (or not) in the satellite receiver (Fig 6.9).

A 'full band' LNB covering the whole Ku band (10.70–12.75 GHz), with a noise figure as low as possible (0.7 to 0.8 dB for the best results), is necesssary, preferably equipped with a

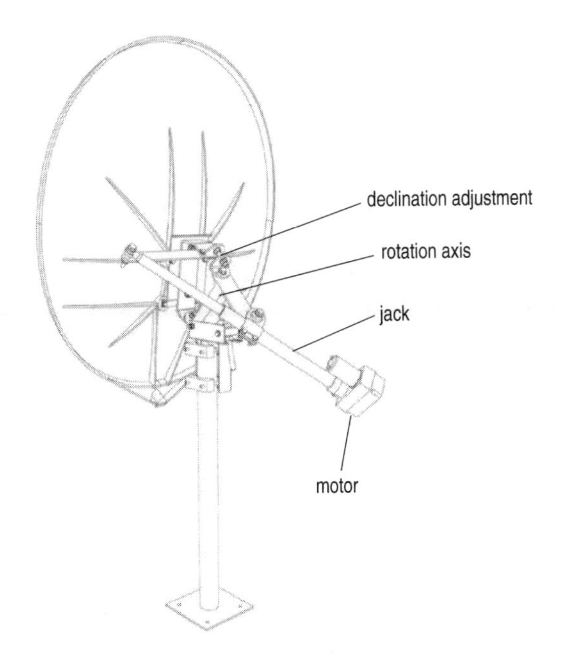

Fig. 6.9 Example of a classical motorized offset antenna (courtesy of Visiosat)

magnetic polarizer rather than a monoblock 'universal' type if the coverage of a large part of the geostationary arc is desired, in order to allow the fine adjustment of the polarization (skew).

In some countries, in order for the user not to require special authorization, the antenna will have to be less than 1 m in diameter, so an offset type or (even better) a Gregorian model should be used owing to its increased efficiency compared with the prime focus models. However, for dimensions larger than 1.20 m, the prime focus models will be practically the only choice. If the 3.7 to 4.2 GHz band is also to be received, the minimum diameter for receiving a large number of channels in acceptable conditions will be 1.50 m, and the noise temperature of the LNB will have to be of the order of 20 to 25 K.

There are some special devices, called *corotors*, which combine source and polarizer and allow coupling LNBs for the C band and Ku band, but these are rather expensive, and not very common, devices. The rather high cost of a C band installation, added to the fact that the C band does not have a large number of TV programmes beyond that of the Ku band, means that its reception is not widespread in Europe and that it is mostly limited to fans of 'DX-TV' and feeds.

Three cables are necessary to control and supply a 'classical' motorized antenna:

1. the coaxial cable, which carries the signal and supplies the LNB;
2. the cable for the antenna motor (two wires for power supply and two wires for control);
3. the cable controlling the polarizer (two or three wires depending on the type).

6.4.2 'One-cable' motorized antenna

It is now possible to find motorized solutions that ensure a good tracking of the geostationary orbit, are easy for the user to install and require only one cable (the coaxial cable) between the antenna and the receiver (or the external positioner) (see Fig. 6.10).

These solutions exploit the coaxial cable for multiple purposes:

- to transmit the received signal;
- to carry the power supply to the LNB and to the motor of the antenna rotator;
- to transmit all commands (antenna rotation, polarization, band, etc.).

Due to the use of the coaxial cable for the motor power supply, the power will be relatively limited (a current of 300 to 400 mA) so that these solutions are generally limited to antennas not exceeding 1 m diameter, and the rotation is relatively slow.

When the positioner is integrated in the receiver, the supply voltage of the motor is generally the same as for the LNB (13 V/18 V), and the supply of the LNB is generally interrupted during antenna rotation. In the case of an external positioner, the motor uses a higher supply voltage, and the voltage for the LNB is reconditioned in the antenna electronic controller.

Since these solutions generally use a universal LNB, there is no skew control; for this reason, and due to the relatively modest dimensions of the antenna, these solutions cannot pretend to cover the complete geostationary arc with the same performance as a classical antenna of larger dimensions; however, they do allow good reception of a number of satellites.

The first models generally used a proprietary control system, based on the same principles as the DiSEqC 1.2 system (modulation of the 22 kHz tone by the control bitstream), to transmit the

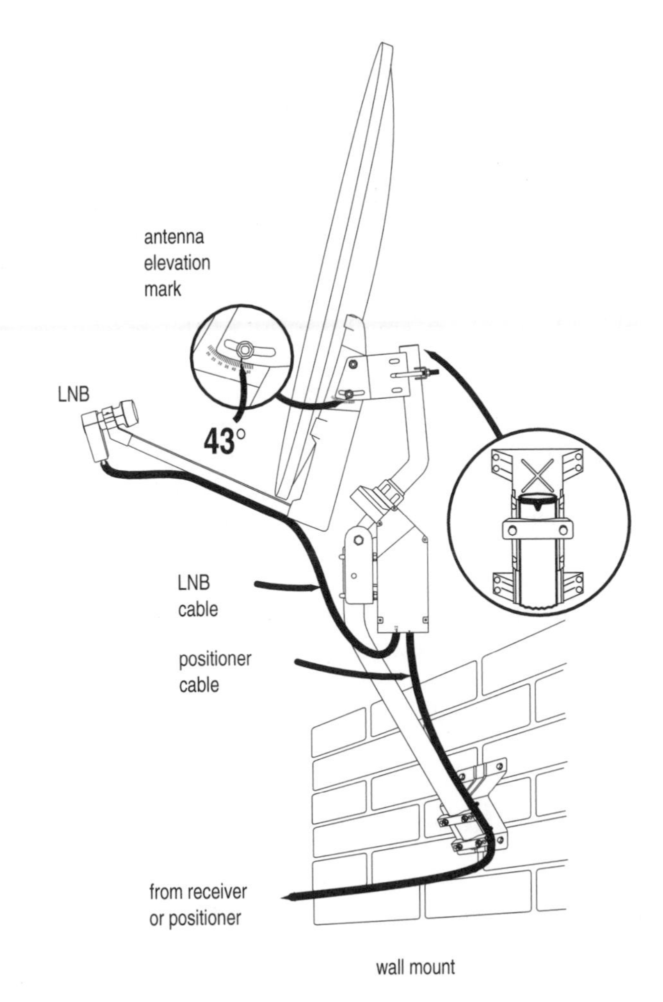

Fig. 6.10 Example of a 'one-cable' motorized antenna (courtesy of Worldsat)

commands to the positioning device. This type of motorized antenna is commercially available either in the form of a kit, including antenna, rotator and receiver with the integrated positioner, or in the form of an add-on kit (rotator plus positioner) adaptable to an existing installation.

The recent introduction of level 1.2 of DiSEqC in some receivers has allowed the appearance of standard motorization kits (the cost of the positioner function being included in the receiver is reduced when it is an original part of the receiver).

6.4.3 'Rotorization'

It is possible to find in the high street so-called cheap 'rotorization' kits, which allow orientation of the azimuth of an antenna, without elevation correction, by means of a rotor controlled and supplied by a separate cable. These systems are, in fact, normally intended to rotate terrestrial antennas for radio or TV. They are not adapted to the reception of satellite since, in addition to the fact that they do not have an elevation correction, which makes them usable only on a very small portion of the geostationary arc around 'full south', their control is generally for a 360° rotation, which makes their adjustment difficult and inaccurate. The net result is that not only is the elevation very approximate, except for one or two particular positions, but also the azimuth, due to the inaccuracy of the adjustment. This is sometimes aggravated due to a rather important 'play' in the rotor, which is perfectly acceptable for a terrestrial antenna, that is much less directionally sensitive than a satellite dish.

All this means this solution is not recommended for a quality satellite installation. It allows a correct reception of only a few orbital positions near each other—an antenna equipped with multiple LNB is technically preferable and sometimes cheaper.

6.5 Antenna with multiple LNB

Multisatellite reception has become very popular in Europe during the last few years (mainly for Astra and Eutelsat), and this is often achieved with a parabolic antenna equipped with multiple LNB (most of the time two LNBs) when the orbital positions of the satellites are near to each other (3 to 10° difference).

In fact, it is possible to receive two or more satellites with the same (offset focus) parabolic antenna, with one of the LNBs being placed at its normal position, and the other being placed at a distance that ensures the required angular offset between the two satellites. In this case, the first satellite is received in a conventional manner and is used as a reference for the adjustment of the antenna position, with the other one(s) being received with a reflection across the symmetry plane of the antenna (see Fig. 6.11).

For this purpose, the second LNB is fixed by means of a device which allows its adjustment relative to the first one, after pointing the antenna at the first satellite (Fig. 6.12). This kind of fixing is mainly used to add a second LNB to an existing antenna.

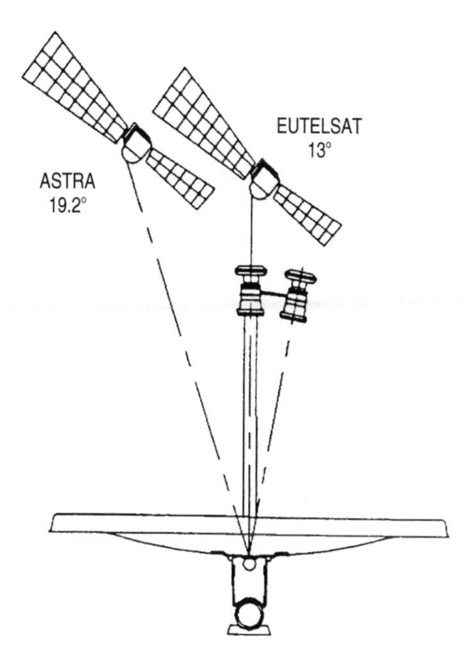

Fig. 6.11 Principle of the reception of two satellites by the same antenna

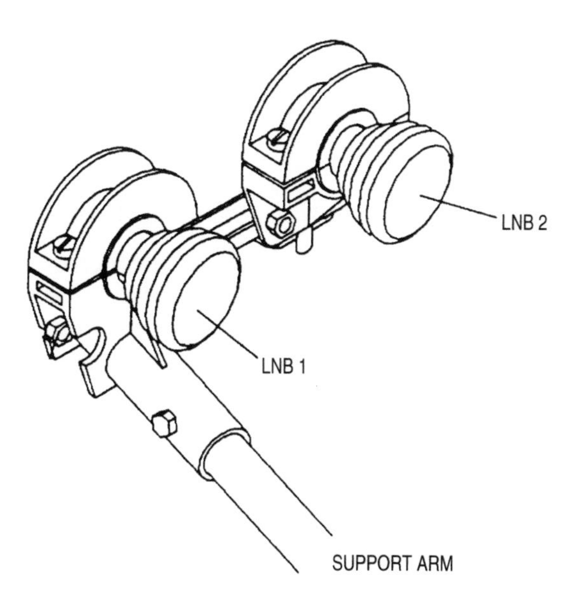

Fig. 6.12 Example of fixing of two LNBs on the same antenna

For new installations, other systems exist where the two LNBs are placed symmetrically in order not to favour one satellite relatively to the other. It is even possible to find some special antennas that can accommodate up to four or five LNBs (beyond these numbers, a motorized antenna is undoubtly the best technical and economic option).

6.6 Collective distribution (SMATV)

The collective distribution of satellite signals is not as simple as one would imagine, and it has become even more complicated in the last few years due to the extension of the Satellite Intermediate Frequency range (SAT-IF from 950 to 2150 MHz), an increase in the number of attractive orbital positions and the coexistence of analogue and digital transmissions. In addition, the problem can be substantially different depending upon whether it is about making a completely new installation or modernizing an existing installation, and on the number of outlets to be served (above 100 outlets, an SMATV installation is considered as a cable network in some countries from the judicial point of view, leading to the question of rights on the delivered TV programmes).

We will examine, but not in too much detail the main possibilities and their respective advantages and drawbacks. The most recent possibility, which offers the same advantages as individual reception, makes good use of the new functionalities offered by DiSEqC.

6.6.1 'Transparent' distribution (switched SAT-IF)

This type of distribution has been designed in order to give to each user (connected by only one coaxial cable to the distribution column) access to all analogue or digital programmes of one or more satellites, using the same satellite receiver, as in individual reception.

In order to give, to each user, transparent access to all channels of the Ku band—divided into two sub-bands (10.7–11.7 GHz and 11.7–12.75 GHz) in both polarizations (horizontal and vertical)—it is neccessary to use an antenna equipped with a 'quattro' LNB for each satellite to be received, with four SAT-IF outputs from 950 to 2150 MHz (one for each polarization of each of the two sub-bands) and to distribute them by means of four cables to the signal distributors (on each floor or storey of a building) which are equipped with a four-position switch for each user output (see Figs 6.13 and 6.14).

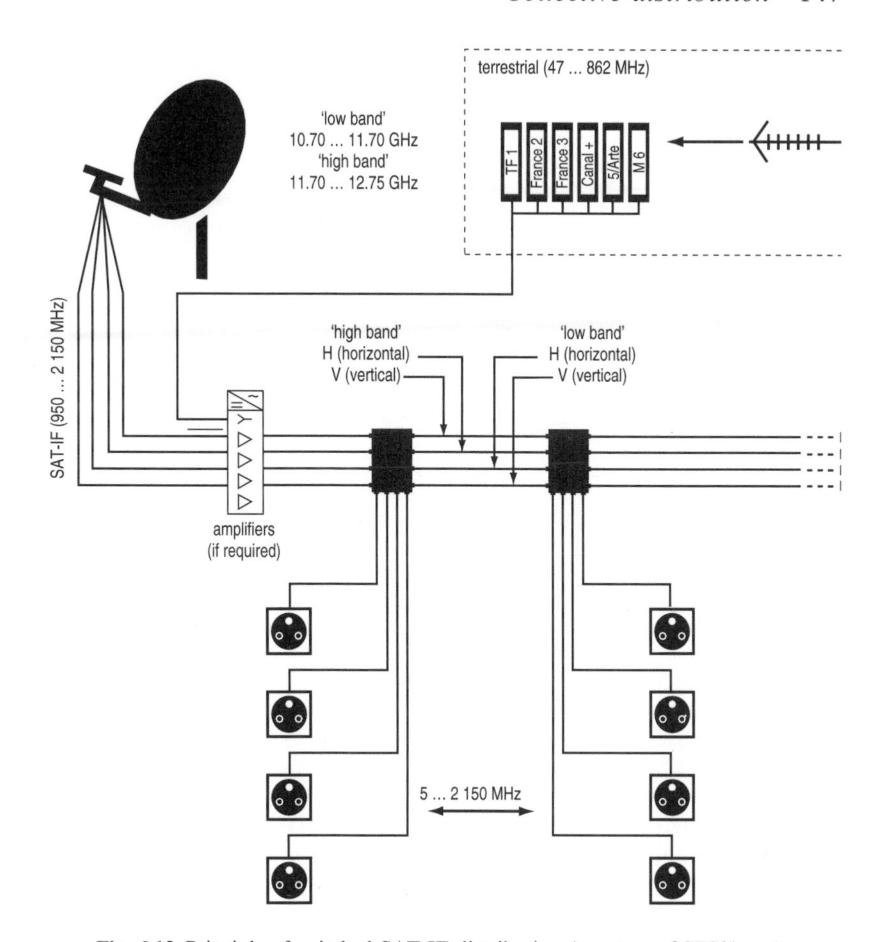

Fig. 6.13 Principle of switched SAT-IF distribution (courtesy of SES/Astra)

Terrestrial TV and FM radio (47 to 860 MHz) are generally coupled by means of a fifth input in the head-end module, and these signals are split by means of filters incorporated in the user outlet, which generally has three outputs (terrestrial TV, FM radio and satellite).

Each user's switching unit is controlled by means of the 13 V/ 18 V voltage (V/H polarization) and the 22 kHz tone (LO/HI band) superimposed on the antenna plug of the receiver. If more than one satellite is to be received (generally two), an extension will be necessary by means of the DiSEqC control (ToneBurst or 1.0).

This distribution system has the advantage of offering the possibility of receiving all transmissions from the selected satellite(s) as an individual antenna, but it has the disadvantage of

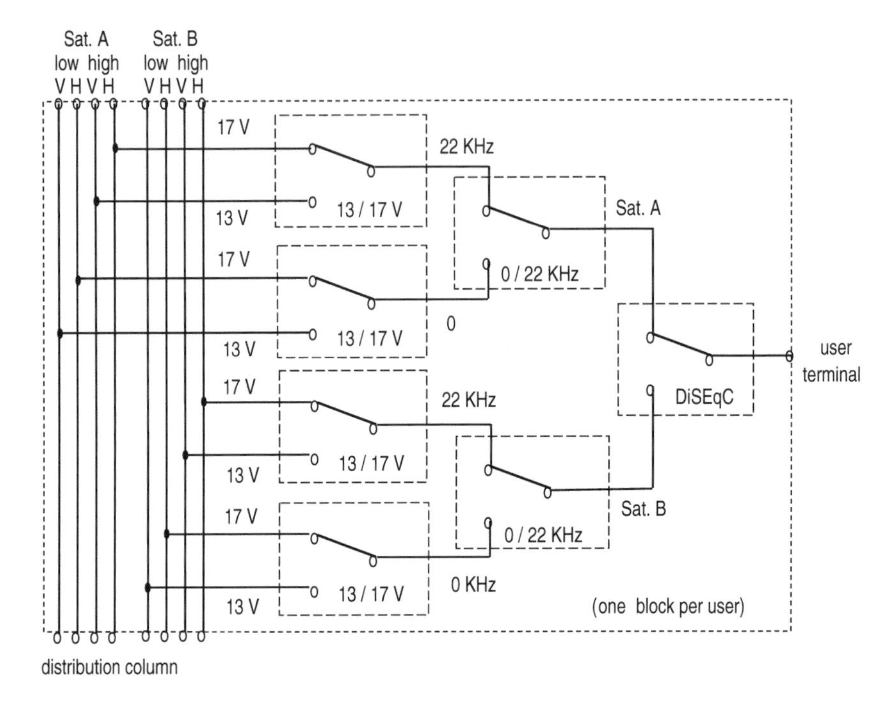

Fig. 6.14 Principle of the user switching unit of a signal distributor for two satellites

requiring four cables per satellite in the distribution column (i.e. eight for receiving ASTRA1 and Eutelsat or ASTRA1 and ASTRA2), which makes it rather costly, and sometimes difficult to install in some buildings.

6.6.2 'Selective' distribution

SAT-IF distribution

In order to avoid high cabling costs, one might want to use only one cable in the distribution column, but this will inevitably lead to making selections of the programmes to be distributed, since the capacity will be limited to around 30 satellite channels (each occupying 40 MHz between 940 and 2150 MHz).

In the most recent installations, wired with 'Ultra Wide Band' (**UWB**) equipment with a bandwidth from 47 to 2150 MHz, the distribution network (cables and distributors) will stay unmodified. In this case the greatest cost will be located in the column head-end equipment, where one frequency transposer will be required per RF channel (transponder carrying analogue or digital signals) to be redistributed in order to create a coherent frequency plan from identical (or near identical) frequency channels coming from multiple satellites, Ku sub-bands and polarizations (Fig. 6.15).

← 950 MHz 2150 MHz →

ALH	1	2	3	4	5	6	7	8	9	10	11	12	13	14	15	16	17	18	19	20	21	22	23	24	25	26	27	28	29	30
ALV	31	32	33	34	35	36	37	38	39	40	41	42	43	44	45	46	47	48	49	50	51	52	53	54	55	56	57	58	59	60
AHH	61	62	63	64	65	66	67	68	69	70	71	72	73	74	75	76	77	78	79	80	81	82	83	84	85	86	87	88	89	90
AHV	91	92	93	94	95	96	97	98	99	100	101	102	103	104	105	106	107	108	109	110	111	112	113	114	115	116	117	118	119	120

ELH	1	2	3	4	5	6	7	8	9	10	11	12	13	14	15	16	17	18	19	20	21	22	23	24	25	26	27	28	29	30
ELV	31	32	33	34	35	36	37	38	39	40	41	42	43	44	45	46	47	48	49	50	51	52	53	54	55	56	57	58	59	60
EHH	61	62	63	64	65	66	67	68	69	70	71	72	73	74	75	76	77	78	79	80	81	82	83	84	85	86	87	88	89	90
EHV	91	92	93	94	95	96	97	98	99	100	101	102	103	104	105	106	107	108	109	110	111	112	113	114	115	116	117	118	119	120

DISTR	2	10	14	23	29	34	39	49	56	63	67	71	75	81	88	95	107	115	109	1	11	24	35	38	50	62	76	85	98	117

Fig. 6.15 Example of selective distribution of two satellites by partial transposition (from 8 × SAT-IF to 1 × SAT-IF)

Legend: ALH = ASTRA, Low band, Horizontal polarization, etc.
EHV = Eutelsat, High band, Vertical polarization, etc.
DISTR = distributed channels (Astra in dark grey, Eutelsat in light grey).
(Channel numbering used here is arbitrary).

The two satellites (e.g. ASTRA and Eutelsat) carry approximately 120 RF channels each (split into two bands and two polarizations) whereas the Ultra Wide Band 'one cable' network can carry only 30 channels between 950 and 2150 MHz in addition to the terrestrial channels. It will therefore only be possible to distribute approximately one programme out of the eight available.

VHF/UHF distribution

In older installations, designed only to distribute terrestrial channels (from 47 to 860 MHz), SAT-IF distribution would generally require a complete replacement of wiring and distributors, which is obviously very costly.

In order to avoid such replacement, it is possible to redistribute a certain number of programmes in the parts of the VHF/UHF band not used by the terrestrial transmissions, but this generally does not allow for redistribution of more than 10 or 12 satellite RF channels (analogue or digital).

For analogue channels, each programme (which occupies a satellite channel of 30 MHz or more) will have to be FM demodulated and remodulated in AM-**VSB** (Amplitude Modulation with Vestigial

Side Band) in a standard 7 or 8 MHz channel. Reception will not require any more equipment other than a standard TV set.

For digital channels, a QPSK satellite channel (DVB-S) of 33 MHz will have to be converted 'transparently' into a QAM cable channel (DVB-C) of 8 MHz in the VHF or UHF band.For this purpose, a QPSK demodulation followed by an error correction (satellite channel decoding) to eliminate transmission errors will be required before a new cable forward error correction (cable channel coding) is applied before 64-QAM modulation. This equipment is called a transmodulator. In this case, the user will need a receiver similar to the models for cable distribution.

Since the bit-rate of a cable channel will be the same as the original satellite channel (of the order of 8 TV programmes per RF channel), 10 to 12 VHF/UHF channels will be able to carry 80 to 100 TV programmes—more or less the complete contents of a satellite 'bouquet'.

A cheaper alternative, consisting of a simple frequency transposition from the SAT-IF band to the VHF/UHF band has been proposed by some suppliers, but in this case the channels keep their original width, which will require a channel spacing of around 40 MHz with guard intervals; this allows only a maximum of four to five RF channels (30 to 40 TV programmes) to be carried, and this is therefore not sufficient to carry any complete satellite bouquet. In this case, the subscriber equipment will be a normal satellite receiver preceded by a frequency transposer which brings the down-converted channels back in the SAT-IF band.

All the above solutions have the drawbacks that they limit to varying degrees the number of receivable channels by the user, and also that they require rather frequent updates when modification in the channel contents occurs (which these days is rather frequent).

6.6.3 Distribution with channel selection by the user in the head-end

DiSEqC 1.1 offers the possibility of the user controlling the tuning of an 'agile' frequency transposer which is dedicated to the user and located in the column head-end, in addition, the user can define the band, polarization and satellite he or she wants to receive.

This allows a new form of distribution of analogue or digital satellite signals using only one cable for the distribution column, at least for installations with fewer than 30 outlets, but still maintaining the possibility of all users receiving all channels available on the antenna system as long as it is equipped with one or more 'quattro' LNBs. For this purpose, a fixed channel (occupying around 40 MHz including guard spaces) among the 30 possible in the 950 to 2150 MHz band, is allocated to each user, who can select the channel to be received by means of his or her transposer (frequency, band, polarization and, possibly even, satellite by means of a DiSEqC 1.1 command, see Fig. 6.5). This channel is always transposed into the same output channel on the distribution network.

This solution, which is still in its early stages, offers the advantage of requiring only one cable in the distribution column without limiting the reception freedom of the user, and without requiring any update in case of frequency changes in the satellite. However, its cost per user is relatively high due to the requirement of one transposer per user in the head-end (the head-end station being modular, however, transposers can be fitted only for interested users at installation, with additional transposers being added on demand for 'latecomers'), and because of relatively sophisticated 'storey' distributors which have to concentrate the DiSEqC messages from the users in order to build a kind of *return channel* to the head-end station.

Other drawbacks of this solution are the limited number of outlets per cable (30) and the fact that it is possible for any user to know which programme his or her neighbour is watching (at

Fig. 6.16 The different DiSEqC logos (courtesy Eutelsat)

least for analogue programmes) as a channel number corresponds to a given user and each user can tune his or her satellite receiver to other channels than their own. This might not be appreciated by everybody.

7 Evolution prospects

In this chapter we look at the foreseeable evolutions of satellite television for the end of this millennium and the first years of the 21st century.

It is certain that digitization will inexorably increase its penetration, first (probably up to 2000 or 2001) by bringing new digital bouquets to all European countries without significantly reducing the number of analogue programmes, which most of the time will be duplicated in digital format ('simulcast'). After this period of analogue and digital 'simulcasting', it is likely that a relatively rapid reduction of the analogue programmes on offer will take place, mainly because the diffusion costs are six to eight times more expensive for analogue than for digital programmes and for a digital transmission of equivalent quality. This will allow an important diffusion capacity to be regained without launching new satellites. The most likely scenario is that, in 2010, analogue transmissions will be a thing of the past.

Digital television will develop very quickly by multiplying the number of available programmes, but also new services will be made possible by the important diffusion capacity available (numerous thematic channels, pay-per-view with short repetition cycle, data diffusion using the 'push' technique which broadcasts data supposed to be of interest to many users, quick Internet access, and so on). In order to profit from the new services without having to use a computer, new terminals will appear which will not only allow TV reception, but also enable 'surfing' the Web at high speed by using a telephone line for the return stream to the 'provider' and either the satellite or the phone line for the downstream, depending on the contents.

Data diffusion services for professional users, which are already open (e.g. ADBS—the Advanced Data Broadcasting System—of Eutelsat, and ASTRANET of SES-Astra), will also develop rapidly.

The two most important orbital positions for Europe (13°E, 19.2°E) are now using the whole Ku band, so new satellite positions will develop, such as the new family of Astra 2 satellites on 28.2°E, which was initially used for the start of digital TV in the United Kingdom (mainly the BSkyB bouquet).

Satellites are also becoming more 'intelligent', being for example able to multiplex many digital SCPC programmes from different origins and to insert the corresponding DVB tables in the satellite itself (e.g. SKYPLEX system of Eutelsat which is operational on Hot Bird 4) in order to form a DVB compliant MCPC channel.

Interactive services will also be possible without using the telephone line as the obligatory return channel; instead, they will use the satellite itself for this purpose, thanks to a return channel in the Ka band (29.5 to 30.0 GHz). This will be possible with antennas not much larger than those used for reception (from 60 to 120 cm depending on the return bit-rate, between 150 kb/s and 2 Mb/s).

The first service of this type foreseen for the time being in Europe is named ARCS (Astra Return Channel System), which is due to start in 1999 on Astra 1H, and will be extended at the end of the year 2000 after the launch of Astra 1K.

What is certain, is that a lot of other innovations, not yet announced, will surprise us!

8 Glossary

8.1 Abbreviations and acronyms

AC-3 – multichannel digital audio system developed by the US company Dolby.

ADR – Astra Digital Radio (digital radio system based on MPEG-2 and using the same subcarriers as analogue sound on the Astra satellites).

AF – Adaptation Field (data field used to adapt the PES to transport packet length).

ASK – Amplitude Shift Keying (digital amplitude modulation with two states).

AU – Access Unit (MPEG coded representation of a presentation unit: picture or sound frame).

AZ-EL – (Azimuth-Elevation) mounting allowing adjustment of a satellite antenna following two axes (horizontal and vertical).

B – Bidirectional (Picture): MPEG picture coded from the preceding and following pictures.

BAT – Bouquet Association Table (optional table of DVB-SI).

BER – Bit Error Rate (number of bits in error relative to the total number of transmitted bits).

BSS – Broadcast Satellite Service (frequency band allocated to direct-to-home transmission, from 11.7 to 12.2 GHz in USA); sometimes used as a synonym of DBS.

CA – Conditional Access (system allowing one to limit and control the access to TV broadcasts).

CAM – Conditional Access Message (specific messages for conditional access: ECM and EMM).

CAT – Conditional Access Table (MPEG-2 table indicating the PID of conditional access packets).

CCIR – Comité Consultatif International de Radiodiffusion (now ITU-R).

CCIR-601 (now ITU-R 601) – recommendation for digitization of video signals (F_s = 13.5 MHz YUV signals in 4:2:2 format).

CCIR-656 (now ITU-R 656) – recommendation for interfacing CCIR-601 signals (most common variant: 8 bits parallel multiplexed YUV format).

CCITT – Comité Consultatif International du Télégraphe et du Téléphone (now ITU-T).

CIF – Common Intermediate Format (360 × 288 @ 30 Hz, for videophone applications).

CNR – Carrier-to-Noise Ratio (or C/N): ratio (in dB) between the received power of the carrier and the noise power in the channel bandwidth.

CSA – Common Scrambling Algorithm (scrambling algorithm specified by DVB).

CVBS – Composite Video Baseband Signal (colour composite video, e.g. NTSC, PAL or SECAM).

D1 – professional digital video recording format in component form (CCIR656).

D2 – professional digital video recording format in composite form (NTSC, PAL or SECAM).

D2MAC – Duobinary Multiplex Analogue Components (hybrid standard used on satellite and cable).

DAB – Digital Audio Broadcasting (European digital audio broadcasting standard).

dB – decibel (unit of relative power): P_R [dB] = 10 log P_2/P_1.

DBS – Direct Broadcast Satellite (satellite in the 11.7–12.5 GHz band reserved for TV broadcast).

dBW – decibel.watt (unit of relative power referring to 1 W): P_R [dBW] = 10 log $P_2/1$.

DC – Direct Current (coefficient of the null frequency in the DCT).

DCT – Discrete Cosine Transform (temporal to frequency transform used in JPEG and MPEG).

DiSEqC™ – Digital Satellite Equipment Control (control protocol allowing digital communication between the set-top box and the antenna by modulating the 22 kHz tone with digital messages).

DPCM – Differential Pulse Code Modulation (digital coding of a value by its difference to the previous one).

DRAM – Dynamic Random Access Memory (read/write memory

requiring a periodic refresh of the information, now often replaced by Synchronous DRAM or SDRAM).

DSR – Digital Satellite Rundfunk (German digital satellite radio system, now obsolete).

DTS – Decoding Time Stamp (indicator of the decoding time of an MPEG access unit).

DVB – Digital Video Broadcasting: name of the group which has specified the European digital TV standard, with three variants, DVB-C (cable), DVB-S (satellite), DVB-T (terrestrial).

DVB-CI – Common Interface (DVB interface for conditional access modules in PCMCIA form).

DVB-SI – System Information (group of tables specified by DVB, additional to MPEG-2 PSI).

DVD – Digital Versatile Disk (unified format of high capacity laser disk, from 4.7 to 19 GBytes).

DXTV – Reception of television at long distance, outside its normal service area (from DX, used in radio amateur language to describe long distance communications).

E_b/N_o – Ratio between the average bit energy E_b and noise density N_o (linked to C/N).

ECM – Entitlement Control Message (first type of conditional access message of the DVB standard).

EIRP – Equivalent Isotropic Radiated Power (of a satellite): the power relative to 1 W (expressed in dBW) which would be required from an isotropic transmitter located at the same distance as the satellite to produce the same power flux density as the satellite at a given point.

EIT – Event Information Table (optional table of DVB-SI indicating a new event).

ELG – European Launching Group (group at the origin of the DVB project in 1991).

EMM – Entitlement Management Message (second type of conditional access message of the DVB standard).

EPG – Electronic Programme Guide (graphical user interface for easier access to DVB programmes).

ES – Elementary Stream (output stream of an MPEG audio or video encoder).

FCC – Federal Communications Commission (regulatory authority for telecommunications in the USA).

FEC – Forward Error Correction (addition of redundancy to a digital signal before transmission allowing correction of errors at the receiving end; synonym: channel coding).

FIFO – First In First Out (type of memory often used as a buffer).

FLASH (EPROM) – non-volatile erasable and reprogrammable memory used for program storage.
FSK – Frequency Shift Keying (digital frequency modulation with two states).
FSS – Fixed Service Satellite (satellite in the 10.7–11.7 GHz or 12.5–12.75 GHz bands originally reserved for telecommunications).
GOP – Group of Pictures (MPEG video layer: succession of pictures starting with an I picture).
G/T – see figure of merit.
HDTV – High Definition Television (TV system with more than 1000 scanning lines).
I – In phase; for QAM, designates the signal modulating the carrier following the 0° axis.
I – Intra (picture); MPEG picture coded without reference to other pictures.
I²C – Inter Integrated Circuits (serial interconnection bus between ICs, developed by Philips).
I²S – Inter Integrated Sound (serial link between digital sound ICs, developed by Philips).
IDTV – Improved Definition Television (TV system with a resolution between standard TV and HDTV).
IEEE1284 – bidirectional high speed parallel interface (enhanced Centronics interface).
IEEE1394 – high speed serial bus (up to 400 Mb/s) which is the future standard for consumer digital A/V links (already used in some digital video recorders and camcorders).
IRD – Integrated Receiver Decoder; popular synonym: set-top box.
ISO7816 – standard defining the mechanical and electrical characteristics of smart cards.
ITU – International Telecommunications Union (world regulatory body for telecommunications, previously CCITT).
JPEG – Joint Photographic Experts Group (standard for fixed pictures compression).
Ku – frequency band used by most TV satellites (10.7 to 12.75 GHz).
LNB or LNC – Low Noise (Block) Converter (situated at the focus of a satellite antenna, converts the incoming Ku band into the SAT-IF band).
MP@ML – Main Profile at Main Level (main video format used by the DVB standard).
MPEG – Motion Pictures Experts Group (group which developed the MPEG-1 and MPEG-2 standards).
MUSE – Japanese High Definition Television system (analogue system with digital assistance).

MUSICAM – Masking Universal Sub-band Integrated Coding And Multiplexing (coding process of MPEG-1 audio, layer 2 used by DAB and DVB).

NICAM – Near Instantaneous Companded Audio Multiplexing (digital sound system mainly used in terrestrial analogue TV using a QPSK modulated subcarrier at 5.85 or 6.55 MHz).

NIT – Network Information Table (optional table of DVB-SI).

NTSC – National Television Standard Committee (colour TV system used in USA and most 60 Hz countries).

OSD – On Screen Display (or On Screen Graphics) display of various information on the TV screen (programme number, tuning status, installation menu etc).

P – Predictive (picture): MPEG picture coded with reference to the preceding I or P picture.

PAL – Phase Alternating Line (colour TV system used in most European and 50 Hz countries).

PAT – Programme Allocation Table (DVB table indicating the PID of the components of a programme).

PC-Card or **PCMCIA** – Personal Computer Memory Card Association (designates the format used for PC extension modules and proposed by DVB for the detachable conditional access modules using the DVB-CI Common Interface).

PCM – Pulse Code Modulation (result of the digitization of an analogue signal).

PCR – Programme Clock Reference (information sent at regular intervals in MPEG-2 to synchronize the decoder's clock to the clock of the programme being decoded).

PES – Packetized Elementary Stream (MPEG elementary stream after packetization).

PID – Packet Identifier (PES identification number in the DVB standard).

PMT – Programme Map Table (DVB table indicating the PID of the PAT of all programmes in a transport multiplex).

PRBS – Pseudo-Random Binary Sequence.

PSI – Programme Specific Information (MPEG-2 mandatory tables: CAT, PAT, PMT).

PTS – Presentation Time Stamp (information indicating the presentation time of a decoded image or sound).

Q – Quadrature; for QAM, designates the signal modulating the carrier following the 90° axis.

QAM – Quadrature Amplitude Modulation (modulation of two orthogonal derivatives of a carrier by two signals).

QEF – Quasi-Error Free (designates a channel with a BER $< 10^{-10}$).

QPSK – Quadrature Phase Shift Keying (phase modulation with four states, equivalent to 4-QAM).
RGB – Red, Green, Blue (the three basic components of a video colour picture).
RISC – Reduced Instruction Set Computer (new microprocessor architecture used among others in recent set-top boxes).
RLC – Run Length Coding (data compression method exploiting repetitions).
RS(204, 188, 8) – abbreviated notation of the Reed–Solomon channel coding used in DVB standards.
RS232 – serial asynchronous communication interface (relatively slow) used in PCs and set-top boxes.
RST – Running Status Table (optional table of DVB-SI informing on the current transmission).
SAT-IF – Satellite Intermediate Frequency (frequency band available at the LNB output, between 950 and 2150 MHz).
SAW – Surface Acoustic Wave filter (bandpass filter used in IF stages; works in ultrasonic mode).
SCART – denotes the Peritel plug (SCART–Syndicat des Constructeurs d'Appareils de Radio et Television, the former name of SIMAVELEC, the French association of A/V equipment manufacturers who defined this plug).
SCR – System Clock Reference (clock reference used in MPEG-1).
SDT – Service Description Table (optional table of DVB-SI).
SECAM – Séquentiel Couleur À Mémoire (colour TV system mainly used in France and in former Eastern Bloc countries).
SIF – Source Intermediate Format (360 × 288 @ 25 Hz or 360 × 240 @ 30 Hz; base of MPEG-1).
SMATV – Satellite Master Antenna Television (satellite signal distribution system using a common antenna for many users).
ST – Stuffing Table (optional table of DVB-SI).
STC – System Time Clock.
TDT – Time and Date Table (optional table of DVB-SI).
TWT – Travelling Wave Tube (power transmission tube used in the satellite transponders).
UWB – Ultra Wide Band (TV signal distribution system with a bandwidth from 47 to 2150 MHz).
VBS – Video Baseband Signal (monochrome composite signal).
VLC – Variable Length Coding (data compression method consisting of coding frequent elements with fewer bits than infrequent ones).
VSB – Vestigial Side Band (AM with one of the two sidebands truncated); used by all analogue TV standards.
WARC – World Administrative Radio Conference (periodic

conference of the ITU which attributes the frequencies of all telecommunication services worldwide).

WSS – Wide Screen Signalling (signalling information on line 23 of the PAL+ signal, also used on standard PAL or SECAM, to indicate the format and other characteristics of the transmission).

8.2 Words and expressions

agile – qualifies a device (e.g. a modulator or a transposer), the frequency of which is programmable, generally by means of a serial bus.

aliasing – disturbance caused by spectrum mixing when sampling a signal with a bandwith exceeding half of the sampling frequency (during an analogue to digital conversion for instance).

apogee – this corresponds to the most distant point from the Earth of an elliptical satellite orbit with the Earth as one focus (used for launching).

azimuth – (of a satellite from a given place) the angle between the horizontal projection of the imaginary line joining the satellite and this point of the Earth and the horizontal North–South line passing through this point.

baseband – original frequency band of an analogue or digital signal before modulation or after demodulation.

bitstream – a continuous stream of bits.

block – in JPEG and MPEG, this designates an 8 × 8 pixel picture to which the DCT is applied.

burst errors – multiple errors occurring in a short time with relatively long periods without error in between.

Cassegrain – parabolic antenna equipped with a secondary hyperbolic mirror reflecting the incoming beam on a point located at the centre of the main mirror.

channel coding – addition of redundancy to a digital signal before transmission allowing correction of errors at the receiving end (synonymous with FEC).

circular – qualifies the polarization of an electromagnetic wave of which the magnetic and electric components 'turn' around the propagation axis at the rate of one turn per wavelength; two circular polarizations exist (left or right) depending on the direction of rotation (this is the opposite of linear polarization).

comb filter – filter used in NTSC or PAL with 'teeth' corresponding to the stripes in the chrominance and luminance spectra for optimum separation of chrominance and luminance.

components video – colour video made of three elementary signals (e.g. RVB or YUV).
composite video – coded colour video using one signal only (NTSC, PAL or SECAM).
constellation – simultaneous display in I/Q coordinates of the points representing all the possible states of a quadrature modulated signal (QAM, QPSK, . . .).
convolutional coding – 'inner' part of the channel coding for satellite transmissions, which increases the redundancy by providing two bitstreams from the original one; it corrects mainly random errors due to noise.
corotor – device, including the source and polarizer, used to couple an LNB for the C band and another for the Ku band on the same parabolic antenna.
cross-polarization (ratio) – the ratio between the received levels of the desired polarization and the orthogonal polarization (it must be as high as possible, not below 20 to 25 dB).
dazzling – (of an antenna) phenomenon occurring (around the equinox periods) when the sun is aligned with the satellite, leading to a saturation of the LNB input stage by solar radiation.
demodulation threshold – point on the curve $S/N_{video} = f(C/N)$ where the curve deviates by 1 dB below the theoretical straight line (it corresponds roughly to the apparition of sparklies on the screen).
downlink – transmission link from the satellite to the (many) receiving earth stations.
eclipse – (of satellite lighting) phenomenon occurring during two periods of 45 days centred around the equinoxes; the satellite is in the shadow of the Earth for some tens of minutes during which all the electrical energy has to be provided by the satellite's batteries.
elevation – (of a satellite from a given place) angle between the imaginary line joining the satellite and a point on the Earth and the horizontal plane passing through this point.
encryption – encoding of information with a key to control its access.
energy dispersal – process aiming to obtain evenly distributed energy after modulation. In analogue TV, it is obtained by superimposing a sawtooth signal onto the video signal and in digital TV, by combining the digital bitstream with a pseudo-random binary sequence (PRBS).
Eurocrypt – conditional access system mainly used with the D2MAC standard.

feeds – occasional satellite transmissions, not intended for direct reception.
field – for an interlaced video signal, this is the group of lines scanned sequentially; the picture (or frame) is made up of two fields (an odd field made of odd lines, and an even field made of even lines with progressive numbering).
figure of merit (*G*/*T*) – ratio between the net gain of the antenna and its total noise temperature; this number, expressed in dB, characterizes the performance of a receiving station, because (at a given place and in given meteorological conditions) the C/N at reception differs from *G*/*T* by a constant.
flicker – disturbing periodic variation of the luminance of a picture when its refresh frequency is too low (below 50 Hz).
flux density – expressed in W/m^2, corresponds to the radiated power from a transmitter per unit area (for an isotropic source, it is obtained by dividing the radiated power by the area of a sphere with a radius equal to the distance from the source to the reception point).
frame (audio) – elementary period of 8 to 12 ms on which the psycho-acoustical coding is performed (corresponds to 12 times 32 PCM samples).
frequency multiplexing – constitution of a composite signal out of different signals, each occupying a part of the frequency band allocated to their transmission.
geostationary – qualifies a satellite in a circular orbit, synchronous with the Earth's rotation (at 35 800 km above the equator), from where it appears at a fixed place for a terrestrial observer.
Gregorian – parabolic antenna equipped with a small secondary parabolic mirror located outside the received beam which is reflected to the LNB placed between this mirror and the main reflector.
group of pictures – in MPEG, qualifies a series of pictures depending from the same Intra picture.
head end – a head end station is the central point of a cable network or MATV system where all the signals to be distributed are received or generated and amplified before being sent over the network.
interlaced scanning – scanning of a picture in two successive fields, one with odd lines, the second with even lines, in order to reduce by a factor of two the bandwidth required for a given resolution and a given refresh rate compared with a progressively scanned picture.
isotropic – qualifies a source radiating uniformly in all directions.
joint_stereo – MPEG audio mode exploiting the redundancy

between left and right channels with two submodes (*MS_stereo*: coding of L+R and L-R; and *intensity_stereo*: coding of common sub-band coefficients for high bands of L and R).

latitude – North–South angular position of a point on the Earth relative to the equator.

layer – in MPEG audio, defines the algorithm used for compression (there are three different layers).

layer – in MPEG video, corresponds to the hierarchical decomposition (from sequence to block).

letterbox – broadcast format used to transmit widescreen films (aspect ratio 16/9 or more) on a standard 4/3 TV screen, leaving two horizontal black stripes at the top and bottom of the picture.

level – in MPEG-2, defines the spatial resolution of the picture to be coded.

line blanking interval – inactive part of the video signal (corresponding to the line flyback period on the screen); it includes the front and back portion of the video signal and the synchronization pulse.

linear – qualifies the polarization of an electromagnetic wave of which the magnetic and electric components have a fixed position; two linear orthogonal polarizations are used (horizontal or vertical) depending on the position of the electric component (this is the opposite of circular polarization).

link budget – calculation allowing determination of the power level received from a satellite using main link parameters (satellite EIRP, attenuations etc) and thus C/N and antenna size.

longitude – East–West angular position of a point on the Earth relative to the Greenwich meridian.

loop-through – refers to the device included in most digital satellite receivers allowing connection of a second (generally analogue) receiver on the same antenna.

macroblock – picture area of 16×16 pixels, made of six blocks: 4 Y, 1 C_b and 1 C_r.

masking – occlusion of the perception of a sound by a more powerful one near in frequency (frequency masking) and/or in time (temporal masking).

Mediaguard – conditional access system of the French company SECA, and used by all digital bouquets of the Canal+ group.

mini DiSEqC – see ToneBurst.

modulation index – in frequency modulation (FM) it is the ratio between the maximum frequency of the modulating signal and the most remote lateral band kept after modulation.

movement estimation – determination of a movement vector allowing an area of a picture to be deduced from a previous picture.

multicrypt – one of the conditional access options in DVB, based on a detachable CA module connected via the 'common interface' DVB-CI.

Nagravision – conditional access system of the Swiss company Kudelski SA, and used by the Spanish digital bouquet Via Digital (this company also developed the Syster analogue conditional access system used mainly by Canal+/Canal satellite analogue programmes).

noise figure – ratio, *N*, between the noise power of a LNB and the noise power delivered at the same temperature (generally 290 K) by a pure resistor equal to the output impedance of the LNB.

noise temperature – (of an antenna) increase in temperature (relative to 290 K) which should be applied to a pure resistor of the same impedance to generate the same noise power as the antenna at 290 K. Noise temperature and noise figure (in dB) are linked by the relation $T\,[\mathrm{K}] = 290\,(10^{NF/10} - 1)$.

offset focus antenna – a parabolic antenna cut in such a way that the LNB is located outside the incoming beam, thus it does not make any shadow on the reflector.

orthogonal sampling – sampling of a video signal by means of a clock locked to the line frequency in order to obtain samples with fixed positions on a rectangular grid.

padding – non-significant bits added to adjust the duration of an audio frame (padding bits) or non-significant stream added to adjust the bit-rate of a bitstream (padding stream).

perigee – corresponds to the the least distant point from the Earth of an elliptical satellite orbit with the Earth as one focus (used for launching).

Peritel – 21 pin audio/video connector (also known as a SCART plug or EUROCONNECTOR) used to interconnect audio-visual equipment (TV, VCR, set-top box etc.).

pixel – (or pel) abbreviation of Picture Element. The smallest element of an imaging or display device. In digital TV, it corresponds to the visual representation of one sample of a digitized picture.

polar axis declination – angle between the reflector of an antenna with polar mounting and its rotation axis (of the order of 6 to 7° at European latitudes).

polar axis elevation – angle between the vertical and the rotation axis of a motorized antenna with polar mounting; it is equal to the latitude of the antenna's location.

polarizer – electromagnetic or electromechanical device located between the source and the LNB used to rotate the polarization by the required angle to make it coincide with the LNB probe.

polar mounting – fixing device used for motorized antennas, of which the rotation axis is parallel to the polar axis, allowing the antennas to follow the geostationary arc during rotation.

pre-emphasis – increase in the amplitude of the high frequencies of an audio or video signal before FM modulation to improve the signal-to-noise ratio (the reverse process at reception is called de-emphasis).

prime focus – qualifies a parabolic antenna in which the LNB is located at the main focus on its symmetry axis, thus within the incoming beam and therefore makes a 'shadow' on the reflector.

profile – in MPEG-2, defines the toolbox used for video encoding.

progressive scanning – scanning of all the lines of a picture in their numerical order in only one frame containing all picture lines (type of scanning used for computer monitors).

psycho-acoustic model – mathematical model of the human aural system behaviour, based among other things on the frequency and temporal masking effects.

puncture – operation consisting of taking only a part of the bit-streams generated by the convolutional coding used in DVB-S and DVB-T to reduce its redundancy, at the expense of a reduced robustness.

quantization – evaluation of a size with a limited number of discrete values (distinct from the quantization step) for instance in an analogue-to-digital conversion or a compression process.

redundancy – indicates that a signal carries the same information many times, thus a part of it can be removed in a compression process.

Reed–Solomon coding – outer part of the DVB channel coding, which adds 16 parity bytes to the 188 byte packets and allows correction of up to 8 bytes per packet; it is denoted RS (204, 188, 8).

return channel – a means to return information from the subscriber's terminal to the service provider generally using the telephone line via a modem (the cable itself is also used in cable TV networks, and satellites will be usable in the near future for satellite television).

reversible coding – coding allowing the exact original information to be recovered by applying the reverse process (this is synonymous with lossless coding; and is the opposite of lossy coding).

roll-off factor – characteristic of the steepness of the filtering applied to a digital signal in order to limit its bandwidth, generally in view of modulation.
sampling – periodic acquisition of the value of an analogue signal, generally with regard to its conversion into a digital number.
scaling factor – in MPEG audio, a 6 bit multiplying factor applied to each sub-band coefficient for the duration of a frame (giving a 128 dB range in 64 values with steps of 2 dB).
scrambling – coding of a signal aimed at preventing its reception without a specific device that ensures the inverse function (descrambling); used in all conditional access systems for pay-TV.
sequence – in MPEG, an uninterrupted series of GOP defined with the same basic parameters.
set-top box – popular name of an Integrated Receiver Decoder (IRD).
simulcast – simultaneous transmission of a programme in two or more standards (e.g. PAL and DVB), generally during a transition period between these standards.
simulcrypt – principle consisting of sending ECM and EMM for more than only one conditional access system for one programme in order to allow reception with different decoder types.
skew – fine tuning of the polarization on a mechanical or magnetic polarizer, required for motorized antennas covering a large part of the geostationary arc
source – device for adapting the LNB input to the beam reflected by the parabolic mirror.
source coding – ensemble of coding operations aimed at reducing the quantity of information delivered by a source (this is synonymous with compression).
sparklies – in analogue satellite TV, this term is used to describe the apparition of white or black dots or 'fishes' on the screen in the case of reception with a C/N too low (below the demodulation threshold).
spectral efficiency – ratio (in bit/s per Hz) between the bit-rate of a bitstream and the bandwidth occupied by the RF signal modulated by this bitstream.
sub-band sample – in MPEG audio, the output signal of one of the 32 sub-band filters (duration: 32 PCM samples, corresponding to 1 ms at 32 kHz sampling rate).
subcarrier – in satellite analogue TV, the carrier located above the video spectrum and FM modulated by an audio signal (linked or not to the video contents).

symbol – in digital transmission, it is the modulating information element. The number of bits per symbol depends on the modulation type (e.g. 2 bit/symbol for QPSK, 6 for 64-QAM).

symbol rate – number of symbols transmitted in a second; in DVB digital satellite TV, the most common symbol rate is 27.5 Msymb/s for a channel of 33 MHz.

table (MPEG-2 PSI and DVB-SI) – descriptive information necessary to access DVB transmissions or to make this access easier.

ToneBurst (or mini DiSEqC) – elementary version of the DiSEqC system allowing only switching between two orbital positions (gives no right to use the DiSEqC logo).

transponder – active part of a telecommunications satellite which rebroadcasts, after frequency change and power amplification, a channel 'uplinked' from the terrestrial station.

uplink – link from the transmitting and controlling Earth station to the satellite.

Viaccess – conditional access system developed by France Telecom, and used among others by the French bouquets TPS and AB Sat.

Videocrypt – conditional access system used mainly by the analogue BSkyB bouquet.

Appendices

A.1 Main satellites receivable in Europe (EIRP, position)

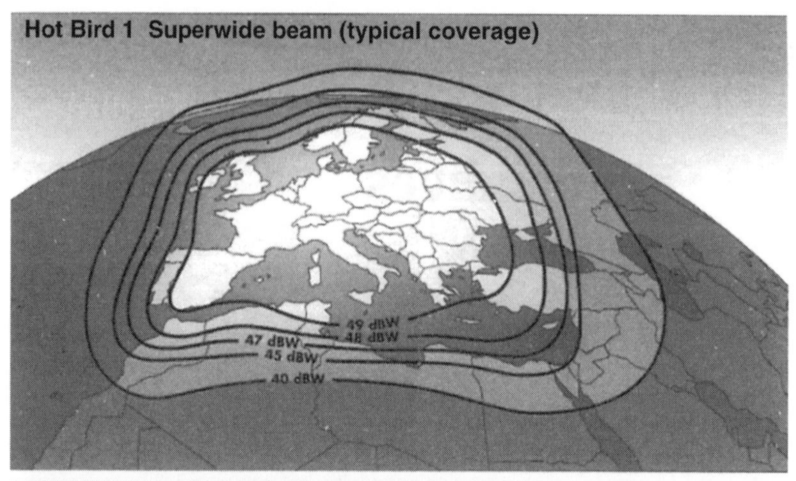

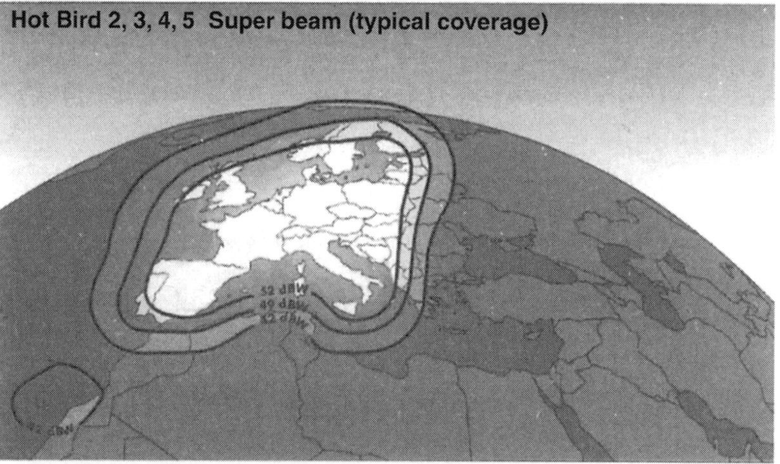

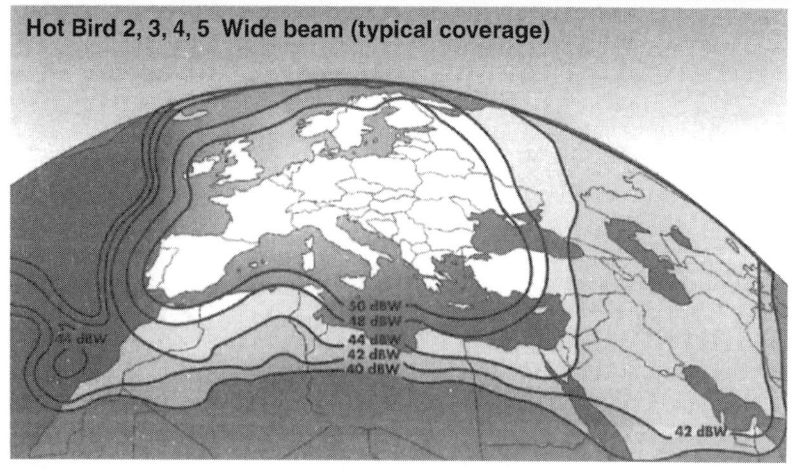

Fig. A1.1 Footprints of the Eutelsat Hot Birds™ satellites (courtesy of Eutelsat)

ASTRA 1C
(10.95 GHz - 11.20 GHz)

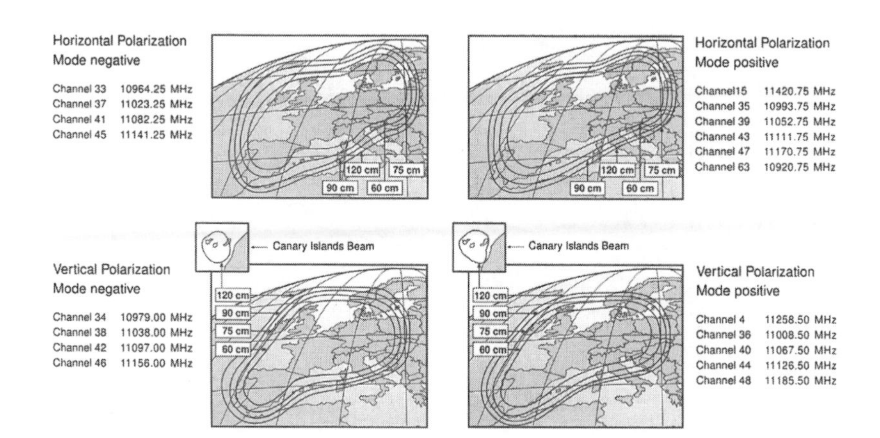

ASTRA 1E
(11.70 GHz - 12.10 GHz)

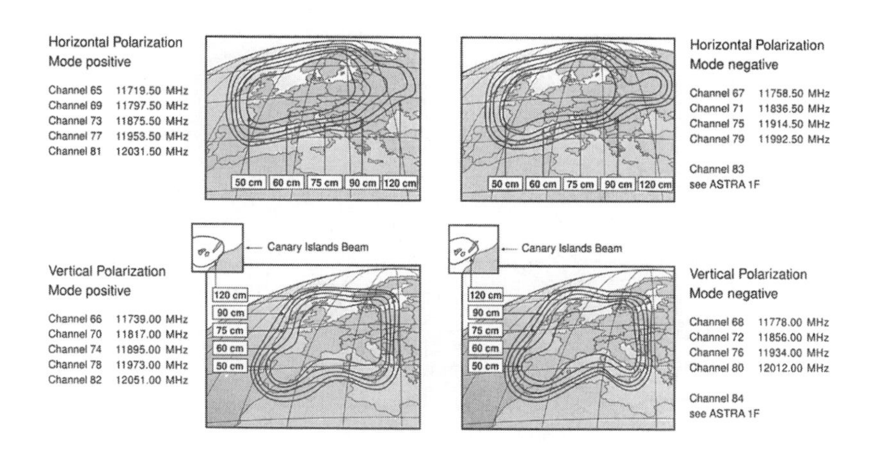

ASTRA 1G
(12.50 GHz - 12.75 GHz)

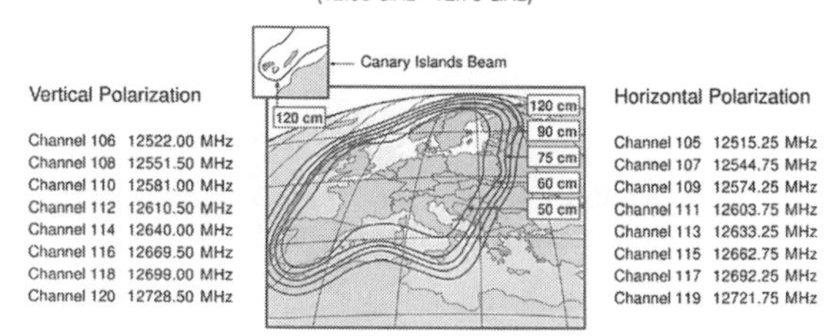

Fig. A1.2 Footprints of the ASTRA 1 satellites (courtesy of SES-Astra)

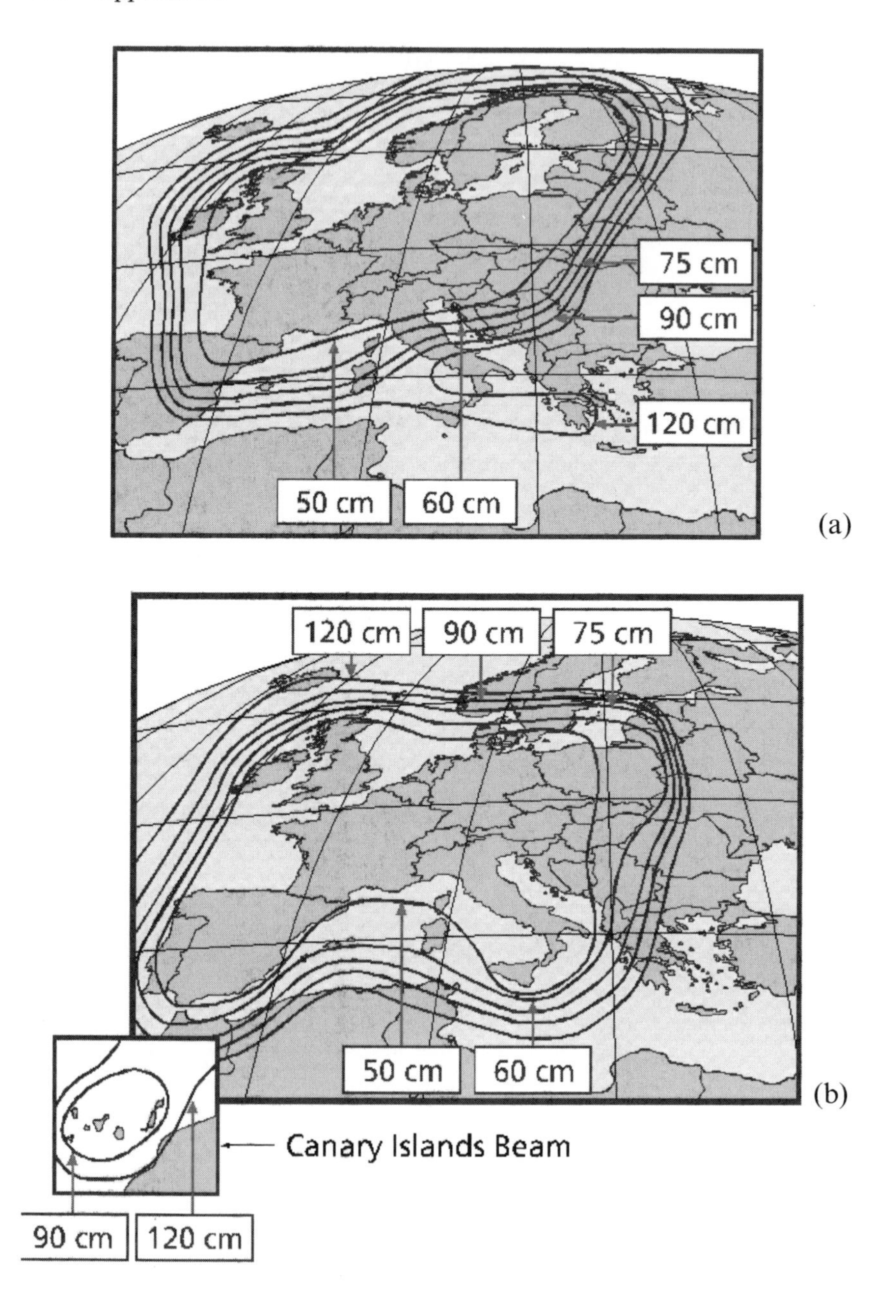

Figs. A1.3(a) (North beam) and **(b)** (South beam)
Footprints of the new satellites ASTRA 2A and 2B (courtesy of SES-Astra)

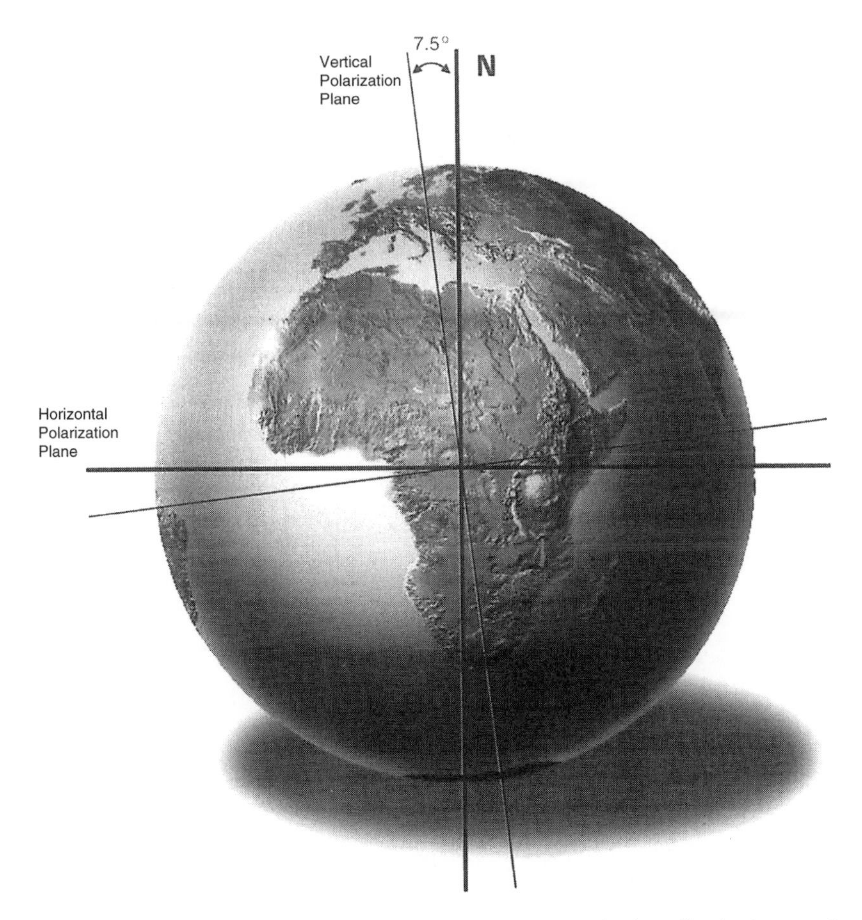

Fig. A1.4 Example of polarization pre-set to minimise the polarization offset in the central part of the target area when the satellite is far from its meridian plane (courtesy of SES-Astra)

```
<!DOCTYPE HTML PUBLIC"-//IETF//DTD HTML//EN">
<html>
 <head>
  <title>Satellite antenna adjustment</title>
  <script>
   var R = 0.1511;           // Earth / Orbit Radius Ratio
   var K = 180 / Math. PI;     // Degrees to Radians Conversion

   function fix(val, len) {  // Function to fix decimal length
     val = val * Math.pow(10, len);
     val = (val < 0) ? Math.ceil(val) : Math.floor(val);
     val = val / Math.pow(10, len);
     return val;
   }

   function evalPostion() {
     for(i=0; !document.data.SatPosition.options[i].selected; i++);

     switch(i) {        // Polarization Offset
      case 3 : D = 9.0; break;
      case 7 : D = 7.5; break;
      case 9 : D = 5.0; break;
      default : D = 0;
     }

     ptn = document.data.SatPosition.options[i].value / K;
     lon = document.data.Longitude.value / K;
     lat  = document.data.Latitude.value / K;

     B = lon - ptn;

     AZ = fix((180 + K * Math.atan(Math.tan(B) * Math.sin(lat))),1);
     if (AZ < 180) document.result.Aximut.value = AZ + "° (east)";
     if (AZ > 180) document.result.Azimut.value = AZ + "° (west";
     if (AZ == 180) document.result.Azimut.value = AZ + "° (full south)";

     EL - fix((K * Math.atan((Math.cos(lat) * Math.cos(B) - R / Math.sqrt(1 -
Math.pow(Math.cos(lat),2) * Math.pow(Math.cos(B),2)))),1);
     if(EL , 2) {
       document.result.Elevation.value = EL + "° " + " Satellite too low!";
     } else {
       document.result.Elevation.value = EL +"°";
     }

     XP = fix((D + (K * Math.atan((Math.sin(B) * Math.sqrt(1 +
Math.pow(Math.sin(Math.atan(Math.sqrt(-1 + (1 / Math.pow(Math.cos(B),2)))) / (6.62 -
Math.cos(B))),2) / Mth.tan(lat)))))),1);
     if (XP < 0) document.result.PolarOffset.value = XP + "° (clockwise seen from sat.)";
     if (XP > 0) document.result.PolarOffset.value = XP + "° (anti-clock seen from sat.)";
     if (XP == 0) document.result.PolarOffset.value = XP + "° (no correction)";
   }
  </script>
 </head>
```

Appendix A1.5 HTML file allowing calculation of azimuth, elevation and polarization offset for the main European satellites depending on the place of reception (continued on next page)

```
<body>
 <h1> Satellite and station data</h1>
Select satellite and input your data, then click <b>RESULTS</b>
<form name="data">
<table><tr><td>
 Choose the Satellite:</td>
<td><select name="SatPosition">
 <option selected value="42.0">Türksat 1C @ 42° East
 <option value="31.0">Türksat 1B @ 31° East
 <option value="28.5">Kopernikus DFS2 @ 28.5° East
 <option value="28.2">Astra 2A@2B @ 28.2° East
 <option value="25.5">Eutelsat IF4 @ 25.5° East
 <option value="23.5">Kopernikus DFS1 @ 23.5° East
 <option value="21.5">Eutelsat IF5 @ 21.5° East
 <option value="19.2">Astra 1A@1G @ 19.2° East
 <option value="16">Eutelsat IIF3 @ 16° East
 <option value="13">HotBird 1@5 @ 13° East
 <option value="10">Eutelsat IIF2 @ 10° East
 <option value="7">Eutelsat IIF4 @ 7° East
 <option value="5">Sirius 2 @ 5° East
 <option value="3">Telecom 2C @ 3° East
 <option value="-1">Intelsat 707 / Thor @ 1° West
 <option value="- 4">Amos 1 @ 4° West
 <option value="-5">Telecom 2B/2D @ 5° West
 <option value="-7">Nilesat 101 @ 7° West
 <option value="- 8">Telecom 2A @ 8° West
 <option value="-12.5">Eutelsat W5 @ 12.5° West
 <option value="-18">Intelsat 705 @ 18° West
 <option value="-21.5">Intelsat Star @ 21.5° West
 <option value="-27.5">Intelsat 605 @ 27.5° West
 <option value="-30">Hispasat @ 30° West
 <option value="-31.5">Intelsat 801 @ 31.5° West
 <option value="-34.5">Intelsat 601 @ 34.5° West
 <option value="-37.5">Orion 1 @ 37.5° West
 <option value="- 43">Panamsat 6 & 3R 43° West
 <option value="- 45">Panamsat @ 45° West
 </select></td></tr>
 <tr><td>Longitude of station (+ = East:<?td><td><input type="text"
name=:Longitude"></td></tr>
 <tr><td>Latitude of station (positive !):<?td><td><input type="text" name"Latitude"></td></tr>
 <tr><td><input type="button" value="RESULTS. . ."onClick="evalPosition();"</td><tr>
 </table>
 </form>
 <hr><h1>Antenna adjustment parameters</h1>
 <form name="result">
 <table>
 <tr><td>Azimuth;</td><td><input type="text" name="Azimut" size="35"></td></tr>
 <tr><td>Elevation:</td><td><input type="text" name="Elevation" size="35"></td></tr>
 <tr><td>Polar Offset:<?td><td><input type+"text" name="PolarOffset" size="35"></td></tr>
 </table>
 </form>
 </body>
</html>
```

Appendix A1.5 HTML file allowing calculation of azimuth, elevation and polarization offset for the main European satellites depending on the place of reception (end)

A.2 Main European satellite channel systems

Table A.0 DBS WARC 77 channels (11.7–12.5 GHz)
(27 MHz channels on a 19.18 MHz grid, circular polarization)

Channel no/ Polarization	Frequency (MHz)	Channel no/ Polarization	Frequency (MHz)	Channel no/ Polarization	Frequency (MHz)	Channel no/ Polarization	Frequency (MHz)
11.7–12.1 GHz				12.1–12.5 GHz			
01 D	11727.48	11 D	11919.28	21 D	12111.08	31 D	12302.88
02 G	11746.66	12 G	11938.46	22 G	12130.26	32 G	12322.06
03 D	11765.84	13 D	11957.64	23 D	12149.44	33 D	12341.24
04 G	11785.02	14 G	11976.82	24 G	12168.62	34 G	12360.42
05 D	11804.20	15 D	11996.00	25 D	12187.80	35 D	12379.60
06 G	11823.38	16 G	12015.18	26 G	12206.98	36 G	12398.78
07 D	11842.56	17 D	12034.36	27 D	12226.16	37 D	12417.96
08 G	11861.74	18 G	12053.54	28 G	12245.34	38 G	12437.14
09 D	11880.92	19 D	12072.72	29 D	12264.52	39 D	12465.91
10 G	11900.10	20 G	12091.90	30 G	12283.70	40 G	12475.50

Table A.1 ASTRA channels in the extended low FSS band (10.7–11.7 GHz) (26 MHz channels on a 14.75 MHz grid, linear polarization)

Channel no/ Polarization	Frequency (MHz)	Channel no/ Polarization	Frequency (MHz)	Channel no/ Polarization	Frequency (MHz)	Channel no/ Polarization	Frequency (MHz)
10.70–10.95 GHz (ASTRA 1D)				11.20–11.45 GHz (ASTRA 1A)			
49H	10714.25	57H	10832.25	01H	11214.25	09H	11332.25
50V	10729.00	58V	10847.00	02V	11229.00	10V	11347.00
51H	10743.75	59H	10861.75	03H	11243.75	11H	11361.75
52V	10758.50	60V	10876.50	04V	11258.50	12V	11376.50
53H	10773.25	61H	10891.25	05H	11273.25	13H	11391.25
54V	10788.00	62V	10906.00	06V	11288.00	14V	11406.00
55H	10802.75	63H	10920.75	07H	11302.75	15H	11420.75
56V	10817.50	64V	10935.50	08V	11317.50	16V	11435.50
10.95–11.20 GHz (ASTRA 1C)				11.45–11.70 GHz (ASTRA 1B)			
33H	10964.25	41H	11082.50	17H	11464.25	25H	11582.50
34V	10979.00	42V	11097.00	18V	11479.00	26V	11597.00
35H	10993.50	43H	11111.75	19H	11493.50	27H	11611.75
36V	11008.50	44V	11126.50	20V	11508.50	28V	11626.50
37H	11023.25	45H	11141.25	21H	11523.25	29H	11641.25
38V	11038.00	46V	11156.00	22V	11538.00	30V	11656.00
39H	11052.75	47H	11170.75	23H	11552.75	31H	11670.75
40V	11067.50	48V	11185.50	24V	11567.50	32V	11685.50

Table A.2 New Eutelsat 'Hot Bird' channels in the extended low FSS band (10.7–11.7 GHz) (33, 36, 47 or 72 MHz, linear polarization)

Channel no/ Polarization	Frequency (MHz)	Channel no/ Polarization	Frequency (MHz)	Channel no/ Polarization	Frequency (MHz)	Channel no/ Polarization	Frequency (MHz)
10.70–10.95 GHz (Hot Bird 4)				11.20–11.45 GHz (Hot Bird 1)			
110V	10718.18	117H	10853.44	01H	11220.75	07H	11345.25
111H	10727.13	118V	10872.62	02V	11241.50	08V	11366.00
112V	10757.54	119H	10891.80	03H	11262.25	09H	11386.75
113H	10775.08	120V	10910.98	04V	11283.00	10V	11407.50
114V	10795.90	121H	10930.16	05H	11303.75	11H	11428.25
115H	10815.08	122V	10949.34	06V	11324.50	12V	11449.00
116V	10834.26						
10.95–11.20 GHz (Hot Bird 5)				11.45–11.70 GHz (Hot Birds 1 and 5)			
				13H	11471.41	155H	11604.10
123H	10971.41	128V	11075.16	14V	11492.16	156V	11623.28
124V	10992.16	129H	11095.91	15H	11512.91	157H	11642.46
125H	11012.91	130V	11116.66	16V	11533.66	158V	11661.64
126V	11033.66	131H	11158.33	153H	11565.74	159H	11680.82
127H	11054.41	132V	11179.08	154V	11584.92		

Table A.3 Telecom channels in the low (11.45–11.7 GHz) and high (12.5–12.75 GHz) FSS bands (36 MHz channels on a 21 MHz grid, linear polarization)

Channel no/ Polarization	Frequency (MHz)	Channel no/ Polarization	Frequency (MHz)	Channel no/ Polarization	Frequency (MHz)	Channel no/ Polarization	Frequency (MHz)
11.45–11.70 GHz (Telecom 2D–5°W)							
12H	11472.00	19V	11535.00	15H	11598.00	22V	11661.00
18V	11493.00	14H	11556.00	21V	11619.00	17H	11682.00
13H	11514.00	20V	11577.00	16H	11640.00		
12.50–12.75 GHz (Telecom 2A–8°W, 2B–5°W, 2C–3°E)							
01V	12522.00	08H	12585.00	04V	12648.00	11H	12711.00
07H	12543.00	03V	12606.00	10H	12669.00	06V	12732.00
02V	12564.00	09H	12627.00	05V	12690.00		

Table A.4 ASTRA channels in the extended DBS band (11.7–12.75 GHz)
(ASTRA 1E/1F and 2A/2B: 33 MHz channels on a 19.50 MHz grid, linear polarization)
(ASTRA 1G and 2B: 26 MHz channels on a 14.75 MHz grid, linear polarization)

Channel no/ Polarization	Frequency (MHz)	Channel no/ Polarization	Frequency (MHz)	Channel no/ Polarization	Frequency (MHz)	Channel no/ Polarization	Frequency (MHz)
11.7– 12.1 GHz (ASTRA 1E and 2A)				12.1–12.5 GHz (ASTRA 1F and 2B)			
65H	11719.50	75H	11914.50	85H	12109.50	95H	12304.50
66V	11739.00	76V	11934.00	86V	12129.00	96V	12324.00
67H	11758.50	77H	11953.50	87H	12148.50	97H	12343.50
68V	11778.00	78V	11973.00	88V	12168.00	98V	12363.00
69H	11797.50	79H	11992.50	89H	12187.50	99H	12382.50
70V	11817.00	80V	12012.00	90V	12207.00	100V	12402.00
71H	11836.50	81H	12031.50	91H	12226.50	101H	12421.50
72V	11856.00	82V	12051.00	92V	12246.00	102V	12441.00
73H	11875.50	83H	12070.50	93H	12265.50	103H	12460.50
74V	11895.00	84V	12090.00	94V	12285.00	104V	12480.00
12.50–12.75 GHz (ASTRA 1G and 2B)							
105H	12515.25	109H	12574.25	113H	12633.25	117H	12692.25
106V	12522.00	110V	12581.00	114V	12640.00	118V	12699.00
107H	12544.75	111H	12603.75	115H	12662.75	119H	12721.75
108V	12551.50	112V	12610.50	116V	12669.50	120V	12728.50

Table A.5 Eutelsat 'Hot Bird' channels in the extended DBS band (11.7–12.75 GHz) (33 MHz on a 19.18 MHz grid, linear polarization)

Channel no/ Polarization	Frequency (MHz)	Channel no/ Polarization	Frequency (MHz)	Channel no/ Polarization	Frequency (MHz)	Channel no/ Polarization	Frequency (MHz)
11.7–12.1 GHz (Hot Bird 2)				12.1–12.5 GHz (Hot Bird 3)			
50V	11727.48	60V	11919.28	70V	12111.08	80V	12302.88
51H	11746.66	61H	11938.46	71H	12130.26	81H	12322.06
52V	11765.84	62V	11957.64	72V	12149.44	82V	12341.24
53H	11785.02	63H	11976.82	73H	12168.62	83H	12360.42
54V	11804.20	64V	11996.00	74V	12187.80	84V	12379.60
55H	11823.38	65H	12015.18	75H	12206.98	85H	12398.78
56V	11842.56	66V	12034.36	76V	12226.16	86V	12417.96
57H	11861.74	67H	12053.54	77H	12245.34	87H	12437.14
58V	11880.92	68V	12072.72	78V	12264.52	88V	12465.91
59H	11900.10	69H	12091.90	79H	12283.70	89H	12475.50
12.50–12.75 GHz (Hot Birds 4 and 5)							
90V	12519.84	93H	12577.38	96V	12634.92	99H	12692.46
91H	12539.02	94V	12596.56	97H	12654.10	100V	12713.28
92V	12558.20	95H	12615.74	98V	12673.28	101H	12730.82

A.3 Example of antenna dimension calculation

The antenna dimension is determined by the minimum C/N to be guaranteed for a certain proportion of the time (generally 99.9%), the minimum EIRP of the satellite(s) to be received, the antenna efficiency and the LNB characteristics (noise figure).

We have seen in Chapter 3 equation (3.1) which relates C/N, G/T and EIRP, and gives:

$$C/N_{min}\,[dB] = G/T_{min} + [EIRP_{min} - (A_{i\,max} + 10\log kB)] \quad (A.3.1)$$

from which one can obtain:

$$G/T_{min}\,[dB] = C/N_{min} - [EIRP_{min} - (A_{i\,max} + 10\log kB)] \quad (A.3.2)$$

It is possible to find the minimum G/T acceptable for the installation by fixing the other parameters:

- the minimal $EIRP_{min}$ of the satellite(s) to be received,
- the worst case atmospheric losses which determine $A_{i\,max}$,
- the bandwidth B of the receiver which determines $10\log kB$,
- the minimum C/N_{min} value required to ensure a good picture (analogue TV) or a bit error rate lower than the correction capability of the forward error correction (digital TV).

It will then be possible to deduce the antenna dimensions if we know:

- the noise figure of the LNB and the antenna noise temperature that determine the system noise temperature T_S and thus allow one to find the net antenna gain G_{net} necessary to guarantee the minimum necessary figure of merit G/T_{min},
- the efficiency of the antenna, which will allow one to find the corresponding theoretical antenna gain G_{theo} of an ideal parabolic antenna and thus its dimensions.

Let us assume an efficiency of 60%, a noise temperature of 55 K for the antenna (excluding LNB) and an LNB noise figure of 1.3 dB.

The LNB temperature T_{LNB} is thus 101.2 K (Table 3.8) and the system temperature T_S=156.2 K.

Hence we get $10\log T_S$ = 21.93 (rounded to 22 dB), which allows us to calculate the necessary net antenna gain G_{net}.

The corresponding theoretical gain G_{theo} (100% efficiency) which allows us to find the antenna dimension (cf. Table 3.1) is obtained in this case by adding 3.7 dB to the net gain net G_{net} (2.2 dB for an antenna efficiency of 60% and 1.5 dB for non-perfect coupling and pointing).

A.3.1 Case of analogue TV

As we have seen before (see section 4.3.4), the picture quality of a TV picture is considered good if the signal-to-noise ratio exceeds 40 dB, which corresponds to a C/N ratio of 11 dB with the modulation and pre-emphasis characteristics used (described in Chapter 4).

Table A.6 gives the minimum figure of merit G/T as a function of the EIRP of the satellite to be received in order to guarantee a minimum C/N of 11 dB for 99.9% of the time ($A_{\text{i max}} = 207.5$ dB) in a bandwidth of 27 MHz in the Ku band (12 GHz). This is obtained by replacing the parameters in (A.3.2) by their values

$$G/T_{(\text{min})}\ [\text{dB}] = 11 - \{\text{EIRP} - [207.5 + 10 \log (1.38 \times 10^{-23} \times 27 \times 10^{6})]\} = 64.2 - \text{EIRP}$$

Table A.6 Figure of merit and antenna dimension as a function of EIRP (analogue TV)

EIRP (dBW)	40	42	44	46	48	50	52	54	56
$G/T = G_{\text{net}}/T_{\text{S}}$	24.2	22.2	20.2	18.2	16.2	14.2	12.2	10.2	8.2
G_{net}	46.2	44.2	42.2	40.2	38.2	36.2	34.2	32.2	30.2
G_{theo}	49.9	47.9	45.9	43.9	41.9	39.9	37.9	35.9	33.9
Φ (cm)$_{\text{min}}$	250	200	160	125	100	80	63	50	40

A.3.2 Case of digital TV

The case of digital TV is not really any more complicated to calculate, but the criteria of good reception are, of course, different.

Taking into account the error correction codes used (see section 5.4.4), reception will be theoretically error free if the bit error rate (BER), after Viterbi, is less than 2×10^{-4} (ETSI specifications), which allows a guaranteed 'quasi-error free' reception (BER of 10^{-10}–10^{-11} after Reed–Solomon decoding).

The $E_{\text{b}}/N_{\text{o}}$ value (see Note A.1) corresponding to this error rate depends on the puncturing rate (or code rate) R_{c} used.

For the most common values of R_c, it is 6.5 dB (R_c = 3/4) and 6.0 dB (R_c = 2/3). This value takes into account an implementation loss of 1.8 dB (see Note A.2). With the used value of $BW/R_S \approx 1.2$ (BW = 33 MHz with R_S = 27.5 Msymb/s or BW = 26 MHz with R_S = 22 Msymb/s), the relation between C/N and E_b/N_o is (Note A.1):

$$C/N = E_b/N_o + 2.2 \text{ [dB]}$$

Table A.7 summarizes these results, where an additional margin of 0.3 dB has been added to take into account possible interferences (for instance due to imperfect cross-polarization rejection and insufficient antenna directivity).

Table A.7 C/N as a function of code rate (digital TV)

Code rate R_c	$E_b/N_{o\ min}$ (BER < 2×10^{-4})	C/N_{min} (BW/R_s = 1.2)	C/N_{min} (with 0.3 dB margin)
3/4	6.5 dB	8.7 dB	9.0 dB
2/3	6.0 dB	8.2 dB	8.5 dB

In practice, the two code rates can co-exist on the same satellite (3/4 being the most common), so we will have to use this value and take for the minimum value, C/N_{min} = 9 dB.

Table A.8 gives the minimum figure of merit G/T as a function of the EIRP of the satellite to be received in order to guarantee a minimum C/N of 9 dB for 99.9% of the time ($A_{i\ max}$ = 207.5 dB) in a bandwidth of 33 MHz in the Ku band (12 GHz). This is obtained by replacing the parameters by their value in (A.3.2):

$$G/T_{(min)} \text{ [dB]} = 9 - \{\text{EIRP} - [207.5 + 10 \log (1.38 \times 10^{-23} \times 33 \times 10^6)]\} = 63.1 - \text{EIRP}$$

Table A.8 Figure of merit and antenna dimension as a function of EIRP (digital TV)

EIRP (dBW)	40	42	44	46	48	50	52	54	56
$G/T = G_{net}/T_S$	23.1	21.1	19.1	17.1	15.1	13.1	11.1	9.1	7.1
G_{net}	45.1	43.1	41.1	39.1	37.1	35.1	33.1	31.1	29.1
G_{theo}	48.8	46.8	44.8	42.8	40.8	38.8	36.8	34.8	32.8
Φ (cm)$_{min}$	220	180	140	110	90	70	55	45	37

One can see that the minimum dimensions are slightly smaller (by 10 to 15% in diameter) than for analogue TV.

Note A.1

The E_b/N_o ratio (average bit energy/noise density) is preferred over the C/N ratio (or CNR) for digital modulations, as it takes into account the number of states of the modulation. There is indeed a relation between these two ratios, which describes the same physical reality. E_b/N_o is linked to E_s/N_o (average symbol energy/noise density) by the relation:

$$E_b/N_o = E_s/N_o \log_2 M$$

where M represents the number of states of the modulation (e.g. four for QPSK, 64 for 64-QAM). E_s/N_o is itself linked to C/N by the relation:

$$E_s/N_o = (\mathrm{C/N}) \times B_{eq} \times T$$

where B_{eq} is the equivalent noise band (≈ channel width BW) and T is the symbol period. From there comes the relation between E_b/N_o and C/N (without FEC):

$$E_b/N_o \approx (C/N) \times BW \times T/\log_2 M$$

C/N and E_b/N_o are normally expressed in dB; the relation then becomes:

$$E_b/N_o\ [\mathrm{dB}] = \mathrm{C/N\ (dB)} + 10 \log [BW \times T/\log_2 (M)]$$

It is more practical to express it as a function of the symbol rate $R_S = 1/T$:

$$E_b/N_o\ [\mathrm{dB}] = \mathrm{C/N\ (dB)} + 10 \log [BW/R_S \,.\, \log_2 (M)]$$

or

$$\mathrm{C/N\ [dB]} = E_b/N_o\ [\mathrm{dB}] + 10 \log [R_S \,.\, \log_2 (M)/BW]$$

In satellite TV, $M = 4$ (QPSK) ; if $BW = 33$ MHz and $R_S = 27.5$ Msymb/s, we then have:

$$\mathrm{C/N} = E_b/N_o + 2.2\ [\mathrm{dB}]$$

Note A.2

These 1.8 dB can be broken down as follows (ETSI specification):

- 0.8 dB due to the implementation loss of the demodulator
- 0.2 dB due to a practical bandwidth, inferior to the bandwidth calculated with the theoretical value of the 'roll-off' (1.2 instead of 1.35)
- 0.8 dB due to the transponder non-linearity.

Index